THE EPIC BOOK OF CRYPTIDS

ETHAN J. HOWARD

A BLIND GUY'S VIEW

Copyright © 2024 by Ethan J. Howard

All rights reserved.

No portion of this book may be reproduced in any form without written permission from the publisher or author, except as permitted by U.S. copyright law.

Dedicated to my wonderful mother. Not because she believes any of these creatures exist (trust me, she does not), but because she is the most supportive, loving, and amazing mother on Earth. I love you dearly, Mom.

Preface

Imagination is one of humanity's most profound gifts. It enables us to dream beyond the known, explore the unknown, and question the boundaries of reality. For centuries, cryptids--mysterious creatures that evade scientific confirmation--have sparked curiosity and wonder. These beings exist in a fascinating space between imagination, science, and mythology, revealing not only the mysteries of our natural world but also the limitless creativity of human storytelling.

Cryptids are not merely products of fanciful tales; they symbolize the enduring human drive to explore and discover. From the remote jungles of the Amazon to the uncharted depths of the ocean, history has shown that the unknown often holds remarkable truths. Creatures once dismissed as legends--the giant squid, the okapi, or the coelacanth--were eventually recognized as real through scientific inquiry. The allure of cryptids lies in this possibility: that somewhere, nature may yet hide extraordinary secrets, waiting for us to uncover them.

The connection between cryptids and mythology further deepens their intrigue. Many creatures revered or feared in myth--dragons, unicorns, and sea serpents--are also regarded as cryptids, rooted in a blend of cultural symbolism, misinterpreted discoveries, and enduring folklore. These stories blur the line between myth and mystery, illustrating our shared desire to make sense of the unexplainable. They remind us that what may seem fantastical to one generation could one day be proven real.

As you turn the pages of the Epic Book of Cryptids, may you be transported into a world where the boundaries between reality and imagination dissolve. This book is more than a collection of enigmatic creatures; it is an invitation to ignite your curiosity, expand your perspective, and celebrate the thrill of discovery. Whether you are a skeptic, a believer, or simply captivated by the unknown, these stories are a testament to humanity's enduring quest for knowledge and wonder.

So, embark on this journey with an open mind and a sense of adventure. May the tales within inspire you to see the world not just as it is, but as it might be--a place full of mystery, discovery, and endless possibility.

Happy reading!

Contents

Fullpage image	XV
1. ALIEN BIG CATS	1
Artistic Depiction	2
2. AMERANTHROPOS	3
Artistic Depiction	4
3. AMAROK	5
Artistic Depiction	6
4. AMMUT	7
Artistic Depiction	8
5. ASWANG	9
Artistic Depiction	10
6. BAKAJIRA	11
Artistic Depiction	12
7. BATURONG	13
Artistic Depiction	14
8. BEAST OF BLADENBORO	15
Artistic Depiction	16
9. BEAST OF DARTMOOR	17
Artistic Depiction	18
10. BEAST OF GÉVAUDAN	19
Artistic Depiction	20
11. BIGFOOT (SASQUATCH)	21
Artistic Depiction	22
12. BLACK SHUCK	23
Artistic Depiction	24
13. BLOBFISH (Verified)	25
Artistic Depiction	26
14. BROSNO DRAGON	27

Artistic Depiction	28
15. BUNYIP	29
Artistic Depiction	30
16. CADBOROSAURUS	31
Artistic Depiction	32
17. CAPE YORK MONSTER	33
Artistic Depiction	34
18. CASSOWARY (Verified)	35
Artistic Depiction	36
19. CHANEKE (Mexico)	37
Artistic Depiction	38
20. CHUPACABRA	39
Artistic Depiction	40
21. CON RIT	41
Artistic Depiction	42
22. CORONADO SPHINX	43
Artistic Depiction	44
23. CRESSIE	45
Artistic Depiction	46
24. DEVIL MONKEY	47
Artistic Depiction	48
25. DOBHAR-CHÚ	49
Artistic Depiction	50
26. DOGMAN	51
Artistic Depiction	52
27. DOVER DEMON	53
Artistic Depiction	54
28. DROP BEAR (S	55
Artistic Depiction	56
29. DUENDE	57
Artistic Depiction	58
30. EYEWITNESS OF GUADALCANAL	59
Artistic Depiction	60
31. FAIRY	61
Artistic Depiction	62

32. FANG SNAKE	63
Artistic Depiction	64
33. FLATWOODS MONSTER	65
Artistic Depiction	66
34. FRESNO NIGHTCRAWLER	67
Artistic Depiction	68
35. GLOUCESTER SEA SERPENT	69
Artistic Depiction	70
36. GOBLIN	71
Artistic Depiction	72
37. GROOTSLANG	73
Artistic Depiction	74
38. HELLHOUND	75
Artistic Depiction	76
39. HODAG	77
Artistic Depiction	78
40. IGOPOGO	79
Artistic Depiction	80
41. INUIT GIANT	81
Artistic Depiction	82
42. JERSEY DEVIL	83
Artistic Depiction	84
43. JUBOKKO	85
Artistic Depiction	86
44. JUMBO BIRDS	87
Artistic Depiction	88
45. KASAI REX	89
Artistic Depiction	90
46. KELPIE	91
Artistic Depiction	92
47. KONGAMATO	93
Artistic Depiction	94
48. KRAMPUS	95
Artistic Depiction	96
49. LA LUZ MALA	97

Artistic Depiction	98
50. LOCH NESS MONSTER	99
Artistic Depiction	100
51. LOPEZ ISLAND BIGFOOT	101
Artistic Depiction	102
52. LOVELAND FROGMAN	103
Artistic Depiction	104
53. LUSCA	105
Artistic Depiction	106
54. MANTICORE	107
Artistic Depiction	108
55. MAPINGUARI	109
Artistic Depiction	110
56. MAROZI	111
Artistic Depiction	112
57. MELON HEADS	113
Artistic Depiction	114
58. MENEHUNE	115
Artistic Depiction	116
59. MERMAID	117
Artistic Depiction	118
60. MINIWAKEE	119
Artistic Depiction	120
61. MOKELE-MBEMBE	121
Artistic Depiction	122
62. MONGOLIAN DEATH WORM	123
Artistic Depiction	124
63. MOTHMAN	125
Artistic Depiction	126
64. MUGWUMP	127
Artistic Depiction	128
65. MUNGO MAN	129
Artistic Depiction	130
66. MUNYANGO	131
Artistic Depiction	132

67. NAITAKA	133
Artistic Depiction	134
68. NAKULA	135
Artistic Depiction	136
69. NANDI BEAR	137
Artistic Depiction	138
70. NGUMA-MONENE	139
Artistic Depiction	140
71. NINKINANKA	141
Artistic Depiction	142
72. NØKKEN	143
Artistic Depiction	144
73. NUE	145
Artistic Depiction	146
74. OGOPOGO	147
Artistic Depiction	148
75. OKONKORO	149
Artistic Depiction	150
76. OWEBRE	151
Artistic Depiction	152
77. PAKTANAK	153
Artistic Depiction	154
78. PESTIMAI	155
Artistic Depiction	156
79. PONTIANAK	157
Artistic Depiction	158
80. POPOBAWA	159
Artistic Depiction	160
81. RAKE	161
Artistic Depiction	162
82. RASPUTIN'S DRAGON	163
Artistic Depiction	164
83. ROPEN	165
Artistic Depiction	166
84. SCORCHING DEATH	167

Artistic Depiction	168
85. SINKHOLE SAM	169
Artistic Depiction	170
86. SKINWALKER	171
Artistic Depiction	172
87. SKUNK APE	173
Artistic Depiction	174
88. SLENDERMAN	175
Artistic Depiction	176
89. SNAKE MAN	177
Artistic Depiction	178
90. TANIWHA	179
Artistic Depiction	180
91. TIKBALANG	181
Artistic Depiction	182
92. TROLL	183
Artistic Depiction	184
93. WENDIGO	185
Artistic Depiction	186
94. YETI	187
Artistic Depiction	188
About the author	189
Full Page Painting	190
Full Page Painting	191

1

ALIEN BIG CATS

(United Kingdom)

Description

Alien Big Cats, also called ABCs, are large, exotic felines such as panthers, leopards, and pumas, reportedly seen in various parts of the United Kingdom. These mysterious creatures are considered "alien" because they are not native to Britain, yet sightings have persisted for decades. Witnesses often describe sleek, muscular cats with dark coats, similar to big cats found in Africa or the Americas. Some reports describe cats up to the size of a Great Dane, with a silent, predatory movement that gives an eerie feel to their encounters.

Origin

The origin of Alien Big Cats in the UK is speculative but rooted in plausible historical circumstances. In the mid-20th century, it was common for exotic animals to be privately owned, especially by wealthy British individuals who kept big cats as pets. When the Dangerous Wild Animals Act of 1976 made private ownership of such animals more challenging and expensive, many owners allegedly released their exotic pets into the wild rather than surrendering them to authorities. This theory suggests that some big cats have survived and even bred, leading to a small but persistent population that continues to evade official documentation.

Lore

Sightings of Alien Big Cats have become modern folklore in Britain, with each region having its own stories and names for the mysterious creatures. Among the most famous are the Beast of Bodmin Moor in Cornwall, the Surrey Puma, and the Beast of Exmoor. The Beast of Bodmin Moor, for example, became the subject of intense media interest in the 1980s after livestock were found mysteriously killed, sparking rumors of a large predator on the loose. Alien Big Cats have captured the public imagination, with countless personal accounts and grainy photographs fueling the mystery.

Additional Facts

- Reports of Alien Big Cats often increase following media coverage of sightings, leading some to believe that the phenomenon may be partially fueled by social influence and local lore.

- Various ecological experts have suggested that, even if Alien Big Cats did exist in the wild, they would likely be solitary and have small populations, making them incredibly hard to spot.

- Despite numerous reports, no verified physical evidence—such as a captured cat or remains—has been presented, leading some skeptics to attribute the sightings to mistaken identities with large domestic cats or even large dogs.

Additional Resources

1. "Mystery Cats of the British Isles" by Merrily Harpur provides an in-depth exploration of ABC sightings and lore across the country.

2

AMERANTHROPOS

(North America)

Description

The Ameranthropos is a cryptid shrouded in mystery, often described as a large, human-like figure spotted across various regions of North America, particularly in remote forests and mountainous areas. Resembling an early hominid, it is typically portrayed as a massive, ape-like creature with distinctly human features, such as a powerful frame, broad shoulders, and a face that blends human and primate traits. Witnesses describe it as having dark, thick fur, a low brow, and deep-set eyes, giving it an ancient and primal aura.

Origin

The Ameranthropos legend has roots in Indigenous folklore, where tales of "wild men" and forest spirits predate European colonization. The name, derived from "American man" in Greek, reflects the belief that it might represent an ancient, uncategorized humanoid species. Over centuries, sightings by trappers, hunters, and explorers have added to the lore, with reports describing a creature that moves with surprising agility, capable of blending seamlessly into dense forest landscapes. Some enthusiasts believe the Ameranthropos could be a surviving remnant of North America's prehistoric past.

Lore

According to folklore, the Ameranthropos is a reclusive creature that avoids human contact, emerging only in the most isolated wilderness areas. In certain Indigenous legends, it is regarded as a guardian spirit of the land, protecting it and punishing those who exploit nature. Tales describe it as highly intelligent, with an awareness of its surroundings that borders on human, and it is said to communicate with forest animals. Unlike more fearsome cryptids, the Ameranthropos is seen as wise and mystical, occasionally guiding lost travelers or leaving markers to warn people of dangers. Some believe it symbolizes the spirit of the North American wilderness itself.

Additional Facts

- The Ameranthropos is sometimes compared to the Sasquatch, though it is often described as more human in appearance, sparking debate over whether they are connected.

- Reports frequently mention unusual behaviors attributed to the Ameranthropos, such as stacking rocks, building rudimentary shelters, or marking territory with tree structures.

- Some researchers suggest it could be a surviving relic of early hominid species like Homo erectus or Neanderthals, although no scientific evidence currently supports this theory.

Additional Resources

1. "Monsters of the Northwoods" by Paul Bartholomew and Robert Bartholomew delves into North American cryptids, including Bigfoot and the Ameranthropos.

3

AMAROK

(Inuit)

Description

The Amarok is a colossal, wolf-like creature in Inuit mythology, often depicted as larger and more powerful than any ordinary wolf. Unlike typical wolves, which move in packs, the Amarok is a solitary hunter. It is said to roam the Arctic wilderness, stalking its prey with remarkable stealth and speed. Accounts describe the Amarok as both awe-inspiring and terrifying, with its muscular build, fierce eyes, and an imposing aura that commands respect and fear.

Origin

The Amarok legend originates from Inuit folklore, with stories passed down through generations of Indigenous Arctic communities. In Inuit culture, the Amarok is typically regarded as a supernatural being rather than a mere animal, representing the wild, untamed nature of the Arctic. The Amarok is sometimes seen as a guardian of balance in the wilderness, punishing those who recklessly disrupt the natural order. As a result, the Amarok has come to symbolize respect for nature and the importance of living harmoniously with the environment.

Lore

According to Inuit legend, the Amarok possesses mystical powers and sometimes acts as a guide or mentor to those lost in the wilderness. In some stories, it challenges individuals seeking strength or bravery, testing them in ways that push them to their limits. One well-known tale tells of a young hunter who wished to gain strength. Upon encountering the Amarok, the hunter was put through rigorous trials. Rather than harming him, the Amarok taught him resilience and courage, allowing the young man to survive and grow wiser. In this way, the Amarok became a powerful symbol in Inuit culture, representing both the dangers and the lessons of the Arctic.

Additional Facts

- Some compare the Amarok to the dire wolf, an extinct prehistoric species, though there is no direct connection. However, the comparison adds to the mystery and ancient feel of the legend.

- Certain Inuit stories depict the Amarok as a protective spirit, safeguarding those who respect the land and punishing those who exploit it.

- As a solitary creature, the Amarok stands apart from other mythological wolves, reinforcing values of independence and the strength found in solitude within Inuit lore.

Additional Resources

1. "The Wolf in the Whale" by Jordanna Max Brodsky is a novel that delves into Inuit mythology, including references to the Amarok and its ties to Inuit spirituality.

2. "Inuit Mythology" by Pamela R. Stern explores the spiritual beliefs and myths of the Inuit people.

4

AMMUT

(Egypt)

Description

Ammut, also known as the "Devourer of the Dead" in ancient Egyptian mythology, is a fearsome figure that embodies divine retribution. This cryptid has a hybrid form, with the head of a crocodile, the forelimbs and torso of a lion, and the hindquarters of a hippopotamus—animals that were seen as the most dangerous predators in ancient Egypt. Positioned at the threshold of the afterlife, Ammut's role is to consume the souls of those deemed unworthy, making it one of the most feared entities in Egyptian lore. With its terrifying features, Ammut served as a stark reminder of the judgment awaiting those who faced the afterlife.

Origin

Ammut's origins are rooted in the mythology surrounding the ancient Egyptian concept of the afterlife, especially the "Weighing of the Heart" ceremony. In this ritual, the god Anubis would weigh the heart of the deceased against the feather of Ma'at, symbolizing truth and justice. If the heart was lighter than the feather, the soul could pass on to the afterlife. However, if the heart was heavy with sin, Ammut would devour it, condemning the soul to eternal damnation. Ammut's role as an enforcer of cosmic order reflects the Egyptian emphasis on balance, justice, and morality.

Lore

Although often seen as a terrifying figure, Ammut plays a crucial role in upholding cosmic justice. By devouring the hearts of the unworthy, Ammut acts as the final deterrent for those who have lived without virtue. In many depictions, Ammut stands near the scales during the Weighing of the Heart, serving as a silent reminder of the consequences of wrongdoing. Unlike other Egyptian deities, Ammut was not worshiped; instead, it served to inspire fear and enforce moral conduct. Representations of Ammut appear throughout ancient Egyptian texts, tombs, and scrolls, often as a cautionary symbol for leading a righteous life.

Additional Facts

- Ammut's name can be translated as "Soul Eater" or "Devourer of Millions," highlighting its role as a consumer of unworthy souls.

- The combination of a crocodile, lion, and hippopotamus represented the fiercest animals known to the Egyptians, making Ammut a powerful composite of these predators.

- Unlike most Egyptian gods and goddesses, Ammut was not considered a deity but rather a force of divine punishment, symbolizing the ultimate consequence of moral failure.

Additional Resources

1. The Egyptian Book of the Dead, translated by E.A. Wallis Budge, which includes depictions of the Weighing of the Heart ceremony and Ammut's role in devouring the unworthy.

5

ASWANG

(Philippines)

Description

The Aswang is a shapeshifting creature from Filipino folklore, notorious for its terrifying nocturnal habits and predation on the vulnerable. Often described as a human by day, the Aswang transforms at night, adopting various forms depending on the legend, such as a dog, bat, or large bird. Known for superhuman speed, strength, and stealth, it can silently stalk its prey. With its pale, elongated features, the Aswang strikes fear into those who know its name, especially as it is rumored to attack during dark, moonless nights.

Origin

The Aswang legend has deep roots in Philippine folklore, with variations across the country's many islands and regions. As one of the most well-known supernatural beings in the Philippines, it has been part of oral traditions for generations. Originally used as a cautionary figure, the Aswang often served to scare children into good behavior. However, some interpretations cast it as a spirit of vengeance or a protector against oppressors. Its form and role vary widely, reflecting the diversity of Filipino culture.

Lore

Aswang tales frequently act as cautionary stories, highlighting societal fears and moral lessons. Many villagers believe the Aswang preys on the sick or helpless, particularly targeting pregnant women and children. It is said to attack with an elongated tongue to drain the life force or blood from its victim. The Aswang's shape-shifting ability, which allows it to appear as a typical person by day, is especially unsettling. This duality—a human by day, a monster by night—adds to its aura of mystery and fear. The Aswang legend often carries warnings about trusting strangers, reflecting a community's wariness toward those who might hide dark intentions.

Additional Facts

- The Aswang is regarded as one of the most feared mythical creatures in the Philippines, and stories about it are widespread across all major regions of the country.

- Some legends include ways to identify an Aswang, such as looking into its eyes, where the reflection will appear upside-down.

- The Aswang's various forms sometimes align it with other supernatural beings, like vampires or witches, depending on the local belief system.

Additional Resources

1. "The Aswang Phenomenon" by Jordan Clark, a documentary examining the mythology and cultural impact of the Aswang in the Philippines.

2. "Creatures of Philippine Lower Mythology" by Maximo D. Ramos, which provides a detailed look at Filipino mythology.

6

BAKAJIRA

(Japan)

Description

The Bakajira is a supernatural sea creature from Japanese folklore, noted for its eerie resemblance to a whale. Unlike ordinary whales, the Bakajira often appears as a ghostly or skeletal figure, with a haunting presence that instills fear in those who witness it. Emerging from the depths on moonless nights, the Bakajira is typically surrounded by an aura of otherworldly mist, enhancing its chilling and ethereal quality. This creature is commonly associated with disaster, often appearing before storms or times of misfortune.

Origin

The legend of the Bakajira originates from Japan's coastal regions, especially among fishing communities. The name "Bakajira" loosely translates to "foolish whale" or "ghost whale," suggesting a being caught between life and death. In some stories, it is seen as the restless spirit of a whale that was hunted and killed but never honored or laid to rest. Occasionally portrayed as a vengeful spirit, the Bakajira is said to haunt the waters where it perished. This folklore highlights the strong connection between Japanese culture and the sea, reflecting the respect and caution shown toward marine life.

Lore

According to traditional lore, the Bakajira appears near fishing boats and coastal villages, often bringing bad luck to those who encounter it. People in coastal communities believe that seeing a Bakajira is a bad omen, signaling poor fishing conditions, impending storms, or other misfortunes. Its spectral form is said to summon unnatural waves that may capsize boats and cause sailors to vanish. Some legends suggest that the Bakajira can summon other sea spirits or ghostly fish to haunt the waters, creating eerie displays in the ocean. Many fishing villages perform rituals, including offerings and prayers, to ward off the Bakajira and appease the spirits of the sea.

Additional Facts

- The Bakajira is associated with yokai (supernatural creatures) and is considered part of Japan's diverse collection of oceanic spirits.

- In certain accounts, it is said to have glowing eyes that pierce through the darkness, adding to its terrifying appearance.

- Bakajira legends emphasize respect for marine life, cautioning against wasteful or disrespectful fishing practices that might disturb the spirits of the sea.

Additional Resources

1. Japandemonium Illustrated by Toriyama Sekien, which includes illustrations and descriptions of various Japanese yokai, including sea spirits similar to the Bakajira.

7

BATURONG

(Malaysia)

Description

The Baturong is a mysterious cryptid from Malaysian folklore, known for its nocturnal nature and its resemblance to a large, flying creature, often compared to a bat. While descriptions vary, the Baturong is commonly depicted with massive, leathery wings, sharp claws, and glowing red eyes that instill fear in those who encounter it. Some stories describe the Baturong with a wolf-like face and sharp fangs, while others suggest a more humanoid or monstrous appearance, adding to its air of dark intrigue. It is often sighted in remote, dense jungles, where locals claim it flies silently, stalking its prey under the cover of night.

Origin

The origins of the Baturong legend are rooted in the rich folklore of Malaysia's indigenous communities, especially among tribes in the state of Sabah on the island of Borneo. Traditionally, the Baturong is viewed as a spirit of the jungle, connected to supernatural forces within the wilderness. Some see it as a guardian of the forest, protecting it from those who disrespect nature, while others regard it as a cursed spirit or demonic figure that preys on travelers or those who stray too far into the jungle. This legend serves as a reminder of the power and mystery that pervade Malaysia's tropical rainforests.

Lore

In Malaysian folklore, the Baturong is seen as both a protector and a harbinger of doom, depending on how individuals approach the jungle. Local stories tell of villagers encountering the creature on full moon nights, when it is said to be most active. The Baturong is believed to silently follow and terrify intruders without being seen, revealing its presence only through its glowing eyes in the darkness. Some tales suggest that those who disrespect the jungle, harm animals, or disturb sacred sites are more likely to encounter the Baturong, adding a moral aspect to its mythos. The creature is both feared and respected, with rituals or offerings sometimes made to ensure safety within its territory.

Additional Facts
- The Baturong is sometimes linked with other regional cryptids, such as the Orang Bunian (mysterious forest spirits), with whom it shares certain spiritual qualities.
- Sightings of the creature are rare, and its description varies, making it one of Malaysian cryptozoology's most elusive figures.
- Some believe the Baturong embodies the spirits of deceased ancestors, adding a layer of ancestral reverence to its lore.

Additional Resources

1. The Encyclopedia of Malaysian Mythology by Mariam Abdul Hamid provides a detailed look at Malaysian cryptids, spirits, and folklore, including the Baturong.

2. Spirits of the Rainforest: Malaysian Mythology and Folklore by Haji Ismail Rahman features tales of creatures like Baturong.

8

BEAST OF BLADENBORO

(North Carolina, USA)

Description

The Beast of Bladenboro is a legendary creature from North Carolina folklore, often described as a large, feline-like animal with dark fur, piercing eyes, and sharp, formidable fangs. Reports frequently describe the beast with a wolf-like face, a long tail, and a crouched, powerful build, giving it a sinister and menacing appearance. Witnesses recount its movements as quick and silent, marked by an unnatural strength and agility that allow it to vanish into the night with ease. The creature is infamous for attacking and killing livestock, and many accounts describe blood-drained animals, adding a mysterious and chilling element to its legend.

Origin

The story of the Beast of Bladenboro began in the early 1950s when residents of Bladenboro, North Carolina, reported a series of attacks on pets and livestock. Initial sightings and attacks in 1953 stirred widespread fear throughout the community. Reports of mutilated animals and eerie encounters led to a local "monster hunt," with armed citizens and hunters scouring the area to find and capture the beast. Although no conclusive evidence was found, the creature's legend grew, becoming an enduring part of Bladenboro's cultural history and captivating cryptid enthusiasts.

Lore

Local lore paints the Beast of Bladenboro as a nocturnal predator, emerging primarily at night to hunt animals and striking fear into the hearts of those who hear its cries. While some believe the creature to be supernatural, others theorize it could be a misidentified big cat, such as a cougar or panther. Over time, the tale has taken on a prominent place in local folklore, with many residents of Bladenboro viewing the creature as a symbol of mystery and fear. The story has also drawn comparisons to other cryptid legends, fueling speculation about what might still inhabit the forests of North Carolina.

Additional Facts

- While theories about the Beast of Bladenboro's true identity abound, none have been confirmed. Suggestions range from a wildcat or cougar to a supernatural being.

- Sightings of the beast decreased following the initial attacks in the 1950s, though occasional reports of large, mysterious animals continue to surface in North Carolina.

- Bladenboro has embraced the legend, even hosting an annual "Beast Fest" to celebrate its mysterious history and cryptid folklore.

Additional Resources

1. The Beast of Bladenboro: The True Story of the Monster That Terrorized a North Carolina Town by Mark Pinsky offers an account of the 1950s events and the creature's impact on local culture.

9

BEAST OF DARTMOOR

(ENGLAND)

Description

The Beast of Dartmoor is a legendary creature reported in the rugged landscapes of Dartmoor, an expansive moorland in Devon, England. This mysterious animal is frequently described as a large, panther-like figure with sleek, dark fur, long limbs, and piercing yellow or green eyes. Sightings often mention its muscular build, long tail, and feline grace, resembling big cats typically found in Africa rather than England. Known for its stealth and speed, the Beast of Dartmoor has sparked both fascination and fear, as it is believed to roam the remote hillsides and woodlands of Dartmoor, especially around dawn and dusk.

Origin

The legend of the Beast of Dartmoor gained widespread attention in the 1980s, when sightings of big, mysterious cats across the UK became increasingly reported. Some believe these creatures were released into the wild after the Dangerous Wild Animals Act of 1976 restricted the ownership of exotic pets. This theory suggests that panthers, leopards, and other large cats might have been freed to avoid legal issues, resulting in isolated populations of these animals adapting to remote areas like Dartmoor. The moor's rugged terrain, dense forests, and low human population make it an ideal habitat for an elusive predator.

Lore

Over time, local lore surrounding the Beast of Dartmoor has grown, with eerie tales from hikers, farmers, and villagers. Some stories describe the beast as a supernatural entity or a spirit guarding the moor, while others connect it to livestock deaths and mysterious tracks in the area. Occasionally, sheep have been found mutilated, leading to fears that the beast hunts within the region. Many locals consider it an enigma—a creature embodying the wild, untamed spirit of the moor. The legend has also been covered in local media and documentaries, adding to its mystique.

Additional Facts
- Sightings of the Beast of Dartmoor are often connected to other big cat sightings across the UK, collectively known as Alien Big Cats (ABCs).
- The region's rugged landscape makes it challenging to confirm or disprove the creature's existence, adding to the intrigue.
- Some naturalists argue that sightings may be exaggerated or the result of mistaken identities involving large domestic cats, dogs, or other animals that inhabit Dartmoor's open land.

Additional Resources

1. Mystery Big Cats by Merrily Harpur, which explores sightings and theories surrounding big cats in the UK, including the Beast of Dartmoor.

10

BEAST OF GÉVAUDAN

(France)

Description

The Beast of Gévaudan is a legendary creature from 18th-century France, often described as a large, wolf-like animal with unusually large teeth, reddish fur, and a muscular build. Eyewitnesses reported that it had a broad chest, an elongated snout, and a long, bushy tail, giving it a terrifying appearance. Its eyes were said to glow with a menacing light, adding to the fear it inspired. Known for its stealth and brutality, the Beast of Gévaudan became infamous for its attacks, which struck fear and fascination across France and beyond.

Origin

The legend of the Beast of Gévaudan began in 1764, when the first attacks were reported in the rural region of Gévaudan, located in what is now the Lozère department of southern France. The creature allegedly targeted both animals and humans, with more than 100 attacks attributed to it over the following years. Concerned by the spread of fear, French authorities, including King Louis XV, took action, sending hunters and soldiers to capture or kill the beast. While several large wolves were killed during these hunts, the attacks continued, fueling the legend of an elusive, possibly supernatural predator. The true identity of the Beast remains a mystery, with theories suggesting it might have been a large wolf, a hyena, or even a hybrid animal.

Lore

The folklore surrounding the Beast of Gévaudan grew with each new reported attack, and stories of the creature's supernatural abilities spread among locals. It was said to be able to leap great distances, withstand gunfire, and vanish into the thick forests of the region. Many believed the Beast was a curse upon the land or even an enchanted creature. Over time, the story of the Beast of Gévaudan became a symbol of resilience and mystery in French culture, becoming one of the most enduring cryptid legends of the country.

Additional Facts

- The Beast of Gévaudan's attacks sparked one of the largest hunts in French history, involving soldiers, hunters, and local volunteers.

- Some modern theories propose that the creature could have been an exotic animal, like a hyena or lion, brought to France and then released or escaped.

- In recent years, the Beast has appeared in French literature, films, and tourism, becoming an integral part of local heritage.

Additional Resources

1. The Beast of Gévaudan: Werewolves, Wolves, and the Mystery of 18th Century France by Jay M. Smith, exploring the historical context and theories surrounding the Beast.

2. The Terror of the French Countryside by Jean-Marc Moriceau, insight into the rural cultural history of France.

11

BIGFOOT (SASQUATCH)

(North America)

Description

Bigfoot, also known as Sasquatch, is a legendary creature said to roam the forests of North America, particularly in the Pacific Northwest. It is described as a towering, ape-like figure standing between 6 and 10 feet tall, with a muscular build covered in dark brown, black, or reddish hair. Eyewitnesses often note its broad shoulders, long arms, and large, human-like feet, which measure up to 24 inches in length. The creature's face is described as somewhat human, with a flat nose, deep-set eyes, and a brow ridge, lending it an eerie, primitive appearance. Bigfoot is also associated with a pungent odor and is known to produce unsettling vocalizations, such as howls, whoops, and whistles. Its mysterious presence has captivated people across the world, blending fear with fascination.

Origin

The legend of Bigfoot originates in Native American folklore, where stories of "wild men" or forest giants have been passed down for generations. The term "Sasquatch" comes from the Halkomelem language of the Coast Salish people in the Pacific Northwest. European settlers later began reporting encounters with a large, hairy creature, but Bigfoot gained widespread attention in 1958 when workers in Bluff Creek, California, discovered large footprints in the area. This sparked media interest, and Bigfoot quickly became a household name. The phenomenon intensified in 1967 with the release of the Patterson-Gimlin film, which allegedly captured footage of a bipedal creature walking through the forest in Northern California. Despite extensive research and numerous sightings, Bigfoot's existence remains unconfirmed, and its true nature continues to be a subject of speculation.

Lore

As sightings increased, Bigfoot became woven into the fabric of North American folklore, often depicted as a mysterious guardian of the wilderness. Stories describe it as a cautious, intelligent creature that avoids human contact, yet leaves behind signs of its presence, such as large footprints, broken tree branches, and eerie sounds echoing through the forests. Some tales imbue Bigfoot with supernatural qualities, suggesting it can vanish at will, evade tracking dogs, or even possess a unique connection to the natural world. Among believers, Bigfoot represents the spirit of the untamed wild, embodying the idea that some mysteries remain hidden in the dense forests. The creature has become an icon of adventure and mystery, sparking curiosity and debates about the unknown.

Additional Facts
- The Pacific Northwest, especially Washington, Oregon, and British Columbia, is known as a hotspot for Bigfoot sightings, with dense forests and mountainous terrain providing ideal, remote habitats.

- Footprint casts are a common piece of evidence collected by enthusiasts, with some showing anatomical details like toe splaying and arch structure, which proponents argue are difficult to fake.

Additional Resources

1. Sasquatch: Legend Meets Science by Dr. Jeffrey Meldrum, a detailed analysis of alleged Bigfoot evidence.

12

BLACK SHUCK

(United Kingdom)

Description

Black Shuck, also known as Old Shuck, is a legendary phantom hound said to haunt the countryside of East Anglia, particularly in the counties of Norfolk and Suffolk. Described as a large, ghostly black dog with shaggy fur, Black Shuck is often depicted as standing anywhere from the size of a large dog to that of a small horse. The creature's most notable feature is its glowing eyes, which are typically red but sometimes green, adding to its sinister aura. In some accounts, Black Shuck is said to have a single, blazing eye in the center of its forehead, giving it an even more otherworldly and menacing appearance. Those who encounter Black Shuck report an overwhelming sense of dread, as the creature is often seen as a death omen.

Origin

The legend of Black Shuck dates back centuries, with some of the earliest accounts rooted in East Anglian folklore and medieval history. The name "Shuck" is thought to derive from the Old English word scucca, meaning "demon" or "evil spirit." It is believed that Norse mythology, brought by Viking settlers, influenced the legend of Black Shuck, particularly the idea of spectral hounds linked to the Wild Hunt—a procession of ghostly figures led by a god or a demonic entity. The most famous Black Shuck sightings occurred on August 4, 1577, during a violent storm. According to legend, Black Shuck appeared at St. Mary's Church in Bungay and later that same day at Holy Trinity Church in Blythburgh. Witnesses reported that the creature burst into the churches, killed several people, and left scorch marks on the doors, known as "the devil's fingerprints." These marks are said to be visible at Blythburgh church to this day.

Lore

The legend of Black Shuck has evolved over centuries, and while the creature is often seen as an omen of death, its role varies across stories. In many accounts, encountering Black Shuck foreshadows death or disaster, with those who see the spectral hound dying shortly after. In some tales, Black Shuck appears to those who are already fated to die, acting more as a herald than a cause of death.

Additional Facts
- The Devil's Fingerprints: Scorch marks on the door of Holy Trinity Church in Blythburgh are believed to have been left by Black Shuck during its 1577 appearance, although skeptics suggest they could be from lightning strikes or other natural causes.

- Shifting Image: While traditionally a death omen, modern interpretations of Black Shuck sometimes portray the hound as a protective spirit, watching over those traveling through dark and lonely paths.

Additional Resources

1. Black Dog Folklore by Mark Norman delves into the legends of spectral black dogs in British folklore, including the tale of Black Shuck.

13

BLOBFISH (Verified)

(Deep Sea)

Description

The blobfish (Psychrolutes marcidus) is a deep-sea creature renowned for its unusual, almost cartoonish appearance, which has led to its reputation as one of the "ugliest" animals in the world. It typically appears as a soft, gelatinous mass with a drooping, human-like face, a large nose, and a downturned mouth. Found at depths between 2,000 and 4,000 feet off the coast of Australia and New Zealand, the blobfish has adapted to the extreme pressures of the deep sea. In its natural habitat, the blobfish has a more streamlined shape, but when brought to the surface, its body decompresses, causing its infamous saggy, amorphous appearance.

Origin

The blobfish was first described in 1926, but its strange appearance did not become widely known until more recent times, when deep-sea exploration began to yield photographs of unusual sea creatures. Part of the Psychrolutidae family, blobfish are related to sculpins and adapted to life on the seafloor. Unlike many fish, they lack a swim bladder, a buoyancy organ that would collapse under the intense pressures of the deep sea. Instead, their gelatinous body has a density slightly lower than seawater, which allows them to float just above the ocean floor. Blobfish remained largely unknown to the general public until images of them surfaced in the early 2000s. Their unusual appearance when brought to the surface quickly made them popular on the internet, leading to memes and a reputation as a "living symbol" of deep-sea oddities.

Lore

While the blobfish itself is not surrounded by traditional folklore, it has become a popular subject in modern culture due to its unique appearance and perceived "ugliness." The creature's droopy, sad-looking face has led it to be affectionately nicknamed "Grumpy Gus" and "Mr. Blobby" in various online communities.

Additional Facts

- The blobfish is found primarily off the coasts of Australia, Tasmania, and New Zealand, residing at depths between 2,000 and 4,000 feet where the pressure is around 100 times greater than at the surface.
- Its body is mostly gelatinous, helping it withstand the extreme pressures of the deep sea and allowing it to float just above the ocean floor with minimal energy.
- Blobfish are bottom-feeders and passive hunters, consuming small crustaceans and other edible matter that drifts past them, relying on the ocean's natural "fall" of organic material.

Additional Resources

1. The Deep: The Extraordinary Creatures of the Abyss by Claire Nouvian provides insight into various deep-sea species, including the blobfish, and their fascinating adaptations.

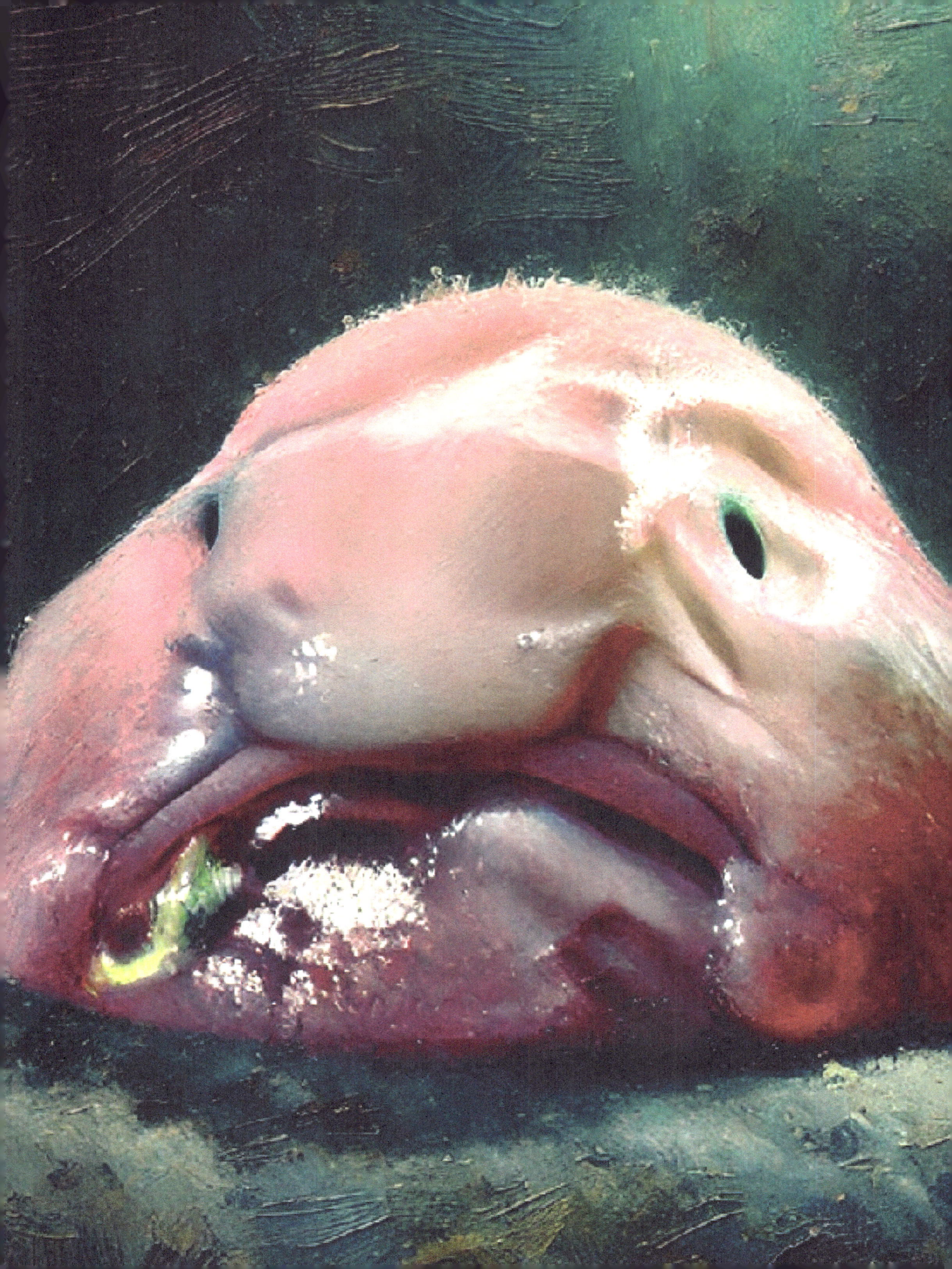

14

BROSNO DRAGON

(Russia)

Description

The Brosno Dragon, also known as "Brosnya," is a mythical creature said to inhabit Lake Brosno, a deep freshwater lake in the Tver region of Russia. Often compared to creatures like the Loch Ness Monster, the Brosno Dragon is described as a large, serpentine figure with dragon-like features. Local folklore describes it as a massive creature with a long, scaly body, a large head, and a powerful tail. Some accounts portray it as having a mouth large enough to swallow boats whole, and it is often said to have eyes that glow beneath the water, adding to its eerie reputation. Legends of the Brosno Dragon have been passed down for centuries, with reported sightings sparking both fear and fascination.

Origin

The origins of the Brosno Dragon legend date back to medieval times, with local tales and stories suggesting that the creature has haunted Lake Brosno for hundreds of years. Stories of the Brosno Dragon gained traction in the 13th century, when Mongol invaders allegedly witnessed a giant creature emerging from the lake and destroying their camp. This incident, recorded in local folklore, supposedly deterred them from advancing further into Russian lands. While these stories lack historical verification, they contribute to the deep-seated mystery surrounding the creature. In more recent centuries, tales of fishermen spotting the dragon and unexplained turbulence in the lake have fueled speculation about the monster's existence.

Lore

The Brosno Dragon has become a symbol of the mysterious, untamed nature of Lake Brosno and its surroundings. In Russian folklore, it is seen as both a fearsome predator and a protective guardian of the lake. Local legends suggest that the creature emerges from the depths to devour anyone who threatens its domain, including boats that come too close. Some stories describe it as an ancient spirit that has watched over the lake for centuries, while others depict it as a remnant of prehistoric times, a creature that somehow survived in the hidden depths of Brosno. Rumors of bubbling waters, strange ripples, and sudden whirlpools are often attributed to the dragon's movements, and sightings of an unknown creature breaking the lake's surface have kept its myth alive. Although skeptics believe the disturbances could be due to underwater gases or other natural phenomena, the Brosno Dragon remains a fixture in Russian folklore.

Additional Facts
- The Brosno Dragon has been compared to other lake monsters around the world, including Scotland's Loch Ness Monster and Canada's Ogopogo.
- Some geologists believe the lake's unusual underwater features, including potential pockets of methane gas, may cause disturbances that resemble the dragon's presence.

Additional Resources

1. Russian Monsters and Mythical Creatures by Aleksandr Ivanov provides detailed accounts of the Brosno Dragon.

15

BUNYIP

(Australia)

Description

The Bunyip is a legendary water-dwelling creature from Australian Aboriginal folklore, often described as a fearsome beast lurking in swamps, rivers, billabongs, and waterholes. Accounts of its appearance vary widely, but common descriptions portray it as having a large, bulky body, long neck, and a head resembling a dog or a seal, with large, glowing eyes and tusks or sharp teeth. Some versions describe the Bunyip as a dark, shadowy figure with a booming roar that echoes across the water, while others depict it as a mysterious, almost shapeless being, blending seamlessly with its murky habitat. The Bunyip is known to instill fear in those who wander near its watery domains, and its elusive nature adds to its mystique. Some accounts describe the Bunyip as bearing a resemblance to a dinosaur, particularly a duck-billed hadrosaur like Muttaburrasaurus, an herbivorous dinosaur that once roamed Australia.

Origin

The Bunyip has been a part of Australian Aboriginal culture for centuries, deeply rooted in the folklore of many Indigenous communities. The word "bunyip" is thought to come from the Wemba-Wemba or Wergaia language of the Aboriginal people in southeastern Australia. Traditionally, the Bunyip is regarded as a guardian of water sources, protecting sacred sites from intruders. Stories of the Bunyip began to gain the attention of European settlers in the 19th century, especially as reports of strange animal remains surfaced. The Bunyip legend soon became widely known, with settlers describing encounters and speculating about its origins.

Lore

In Aboriginal lore, the Bunyip is often seen as a protector of the land and water, punishing those who violate sacred areas or fail to respect the natural world. Tales warn against approaching the Bunyip's territory, as the creature is believed to drag intruders into the depths, never to be seen again. Many stories describe the Bunyip as having supernatural abilities, including shape-shifting, making it nearly impossible to track. The creature's roar is said to be so powerful it can paralyze those who hear it, adding a supernatural element to its legend. In some regions, the Bunyip is thought to be a benevolent spirit rather than a monster, seen as a guardian figure. Over time, the Bunyip has become an iconic figure in Australian folklore, symbolizing the mysterious dangers of the untamed wilderness.

Additional Facts

- Bunyip sightings have been reported across Australia, particularly in the southeastern regions, where wetlands and billabongs are common.

- In the 19th century, the discovery of unusual animal bones in Australian swamps and riverbeds was often attributed to the Bunyip, sparking public fascination and debate.

Additional Resources

1. The Bunyip and Other Mythical Beasts by Robert Holden delves into Australian folklore and the Bunyip.

16

CADBOROSAURUS

(Canada)

Description

Cadborosaurus, often referred to as "Caddy," is a cryptid from Canadian folklore said to inhabit the coastal waters of the North Pacific, particularly around British Columbia. Named after Cadboro Bay in Victoria, where numerous sightings have been reported, Cadborosaurus is described as a long, serpentine creature resembling a sea serpent. Witnesses describe it as having a horse-like head with large eyes, a long neck, and humps along its body, which can range from 10 to 50 feet in length. Some accounts suggest that Caddy has flippers, a fish-like tail, and a slender, undulating body that allows it to move gracefully through the water, creating a ripple effect. Its appearance has led to comparisons with mythical sea serpents, sparking curiosity and fear among locals and tourists alike.

Origin

The legend of Cadborosaurus has deep roots in Indigenous Canadian folklore, where the creature is referred to by various names and believed to be a guardian or spirit of the ocean. The name "Cadborosaurus" originated in the 1930s after reports of a strange creature spotted in Cadboro Bay gained media attention. Since then, Caddy has been spotted along the British Columbian coast and as far south as California and as far north as Alaska, with reports spanning decades. Sightings increased after World War II, with numerous fishermen, sailors, and tourists describing encounters with a strange, serpentine creature in the Pacific waters. Despite the many sightings, no conclusive evidence of Caddy's existence has been found, though occasional bones, carcasses, and unusual remains washed ashore have fueled speculation.

Lore

Cadborosaurus has become a beloved yet mysterious figure in Canadian cryptozoology. Local lore casts Caddy as a cautious creature, rarely seen by humans and often disappearing before it can be closely observed. Indigenous tales describe it as a supernatural being with protective qualities, keeping a watchful eye over the waters and preserving the natural balance of the ocean. In modern times, Caddy is often depicted as a shy but benevolent creature, occasionally glimpsed by lucky observers. Fishermen and boaters tell stories of a creature that moves silently through the water, its undulating body creating distinctive ripples. Some tales suggest that Caddy has avoided capture due to its intelligence and ability to evade human contact.

Additional Facts
- Cadborosaurus sightings are most common off the coast of British Columbia, particularly around Cadboro Bay, where the creature is believed to frequent.
- Occasional reports of strange carcasses washing ashore, known as "globsters," have been linked to Caddy, though these remains are often found to be decomposing whales or other known marine animals.

Additional Resources

1. In the Wake of Cadborosaurus: The Scientific Evidence for an Elusive Sea Serpent by Dr. Paul H. LeBlond.

17

CAPE YORK MONSTER

(Australia)

Description

The Cape York Monster is a cryptid rumored to inhabit the remote regions of Cape York Peninsula in Queensland, Australia. Descriptions of the creature vary, but it is often depicted as a large, bipedal figure with a muscular build, thick fur, and long, sharp claws. Witnesses report seeing glowing eyes and a broad, ape-like face with a pronounced brow ridge and large, sharp teeth. Its size is said to rival that of a large bear, and it reportedly emits a foul, musky odor. The creature's elusive nature and the vast, dense rainforest of Cape York have made it difficult to study, yet its mysterious presence has intrigued locals and visitors alike.

Origin

The legend of the Cape York Monster is deeply rooted in Indigenous Australian folklore, where it is sometimes referred to as a "hairy man" or "yowie." Traditional stories among Aboriginal communities describe a powerful, supernatural being that roams the wilderness, often as a guardian spirit of the land. European settlers began reporting encounters in the early 20th century, with descriptions aligning with Indigenous tales of a large, fearsome creature. Sightings have persisted into modern times, with travelers, hunters, and locals recounting their experiences with the monster. Despite limited evidence, the Cape York Monster remains a focal point of local lore, sparking interest in cryptozoological circles.

Lore

The Cape York Monster is seen as both a harbinger of danger and a symbol of the wilderness's untamed spirit. Indigenous tales describe it as a creature to be respected and avoided, with some stories suggesting it has supernatural abilities, such as disappearing into thin air or evading any attempt to track it. In more recent stories, the creature is often described as a solitary, nocturnal beast that watches humans from a distance, occasionally letting out deep growls or howls. Some tales warn that encountering the Cape York Monster is a sign of bad luck or an omen of misfortune, and those who venture too far into its territory may face harm. Despite its frightening reputation, the Cape York Monster is also viewed as a protector of the wild, embodying the mystery and majesty of Australia's remote landscapes.

Additional Facts

- Cape York Monster sightings are most frequently reported in the rugged, forested areas of Queensland's Cape York Peninsula, particularly near remote rivers and caves. It is often compared to Australia's yowie, though some locals believe it to be a separate creature unique to Cape York.

- Sightings typically occur at dusk or during the night, as the monster is believed to be a nocturnal creature. Hunters and campers report hearing deep growls and experiencing an intense feeling of being watched, even without seeing the creature directly.

Additional Resources

1. Outback Horrors by Samuel Haig explores various cryptids and legends from the Australian wilderness.

18

CASSOWARY (Verified)

(Australia)

Description

The cassowary is a large, flightless bird native to the tropical rainforests of northern Australia, New Guinea, and nearby islands. Known for its striking appearance, the cassowary has glossy black feathers, a bright blue neck, and a red wattle, creating a vivid display of color. Its most distinctive feature is a helmet-like casque on its head, which it uses to navigate through dense foliage. Cassowaries can reach up to six feet in height and weigh as much as 130 pounds, making them one of the heaviest bird species. They have powerful legs and sharp, dagger-like claws on their inner toes, capable of delivering fatal kicks when threatened. Known for their solitary and elusive nature, cassowaries are both respected and feared in their natural habitats.

Origin

The cassowary has long been known to Indigenous Australian cultures, where it is regarded with reverence and caution. Traditional tales often depict the cassowary as a fierce and mystical creature of the forest, embodying the untamed spirit of the jungle. European explorers first encountered the bird in the 19th century and were both fascinated and intimidated by its formidable appearance. The bird's reputation as "the world's most dangerous bird" is due to its powerful legs and sharp claws, which it uses defensively. Despite their potential danger, cassowaries play an essential ecological role, dispersing seeds of various rainforest plants and contributing to the health of their ecosystems.

Lore

In Indigenous Australian folklore, the cassowary is often portrayed as a guardian of the forest. Stories describe encounters with this bird as rare but powerful, with some traditions viewing the cassowary as a symbol of protection. In regions where cassowaries are common, locals maintain a respectful distance, as the birds are known to become aggressive if approached too closely, especially during breeding season or when protecting chicks. Modern lore around the cassowary emphasizes its status as a living relic of prehistory, connecting it to dinosaurs due to its appearance and strength. The bird's reputation has become popularized, with its aggressive behavior contributing to its mystique, yet many people regard the cassowary with a mix of fear and admiration.

Additional Facts
- Cassowaries can run up to 31 miles per hour, making them one of the fastest large birds on land.
- The bird's casque is believed to help it navigate dense underbrush, possibly also serving in temperature regulation or amplifying low-frequency calls.

Additional Resources

1. Birds of Australia by Ken Simpson and Nicolas Day offers a comprehensive look at cassowaries.

19

CHANEKE (Mexico)

(Mexico)

Description

The Chaneke, sometimes known as "chaneque" or "chaneques," is a supernatural creature from Mexican folklore, particularly within Indigenous cultures of Veracruz, Oaxaca, and other regions. Described as small, mischievous beings, Chanekes typically have childlike appearances with unusual features, such as elongated limbs, large eyes, or animal-like characteristics. These diminutive creatures are said to inhabit dense forests, rivers, and caves, where they guard the natural world and occasionally interact with humans. Chanekes are often depicted as playful tricksters with the ability to confuse or disorient people who wander too close to their hidden domains.

Origin

The Chaneke legend has roots in ancient Mesoamerican beliefs, where such beings were revered as nature spirits with protective roles. In Nahuatl, the word "chaneque" translates to "those who inhabit dangerous places," reflecting their association with the untamed wilderness. Indigenous communities saw Chanekes as guardians of sacred areas, responsible for preserving the balance of nature. When the Spanish arrived, they recorded stories of these mythical beings, blending Indigenous beliefs with Spanish folklore. The Chaneke legend has endured for centuries, with sightings and stories passed down through generations, each adding layers to the creature's mysterious persona.

Lore

Chanekes are commonly regarded as tricksters, capable of causing confusion or even temporary illness to humans who disrespect nature or trespass on their territory. Local tales often portray them as protectors of forests, rivers, and other natural sites, rewarding those who show respect and punishing those who harm the environment. Many believe that if someone disrespects the land, Chanekes may lead them astray, causing them to become lost for hours or even days. Conversely, those who offer gifts or prayers to the Chanekes are sometimes rewarded with good fortune. In certain stories, Chanekes also play with children, engaging in games or giving them small treasures. These stories have transformed the Chaneke into both a feared and respected figure in Mexican folklore.

Additional Facts
- Sightings of Chanekes are most commonly reported in dense, isolated areas of Mexico, especially near rivers and remote villages.
- Some stories describe Chanekes as shape-shifters, capable of transforming into animals or even blending into their natural surroundings to remain unseen.

Additional Resources

1. Folklore of Mexico by Elena Gonzales contains stories of Chanekes and other Indigenous Mexican legends.

20

CHUPACABRA

(Puerto Rico)

Description

The Chupacabra, meaning "goat-sucker" in Spanish, is a cryptid reportedly seen in Puerto Rico and other parts of Latin America. Known for its alleged attacks on livestock, particularly goats, the Chupacabra is described as a small to medium-sized creature with reptilian or canine features, large red eyes, sharp teeth, and spines running down its back. It is often depicted as having leathery, grayish-green skin, a row of spines from neck to tail, and a hunched posture. Witnesses claim it has an unusual method of killing by draining its victims' blood, often leaving behind puncture marks on the animals.

Origin

The Chupacabra legend originated in Puerto Rico in the mid-1990s after reports of mysterious livestock deaths began circulating. The first widely publicized sighting occurred in 1995, when a local farmer found eight sheep dead with puncture marks on their necks and their blood reportedly drained. Soon after, more sightings were reported, and descriptions of a bizarre creature began emerging. The stories quickly spread, fueled by media coverage, turning the Chupacabra into a global phenomenon. Some researchers speculate that the Chupacabra legend was influenced by earlier folklore involving vampire-like creatures, while others believe it may stem from natural predation by known animals.

Lore

The Chupacabra has become an iconic figure in modern Latin American folklore, with numerous theories about its origins and motivations. In Puerto Rican culture, it is often seen as a supernatural or extraterrestrial creature, linked to government experiments, alien sightings, or even genetic mutations. Some locals believe the Chupacabra is a manifestation of darker forces, preying on animals and occasionally instilling fear in rural communities. Over time, the Chupacabra legend has evolved to include tales of the creature's elusive nature and its seeming ability to vanish without a trace. Despite a lack of concrete evidence, stories of the Chupacabra persist, with sightings continuing to be reported in Puerto Rico, Mexico, and the southern United States. The creature's mythos has even expanded to include its portrayal as a shape-shifter or spirit capable of taking on different forms to evade capture.

Additional Facts
- The creature's reputation has led to numerous depictions in media, including films, books, and TV shows, turning it into a cultural icon within paranormal and cryptid circles.
- Despite numerous investigations, no confirmed evidence of the Chupacabra has been found, making it one of the most mysterious cryptids in contemporary folklore.

Additional Resources

1. Tracking the Chupacabra: The Vampire Beast in Fact, Fiction, and Folklore by Benjamin Radford examines the origins of the Chupacabra legend, including a detailed analysis of sightings and theories.

21

CON RIT

(Vietnam)

Description

Con Rit, also known as the "Giant Sea Centipede," is a cryptid from Vietnamese folklore believed to inhabit the coastal waters of Southeast Asia. Descriptions of Con Rit vary, but it is generally depicted as a massive, segmented creature resembling an oversized centipede or millipede, with a long, armored body stretching up to 60 feet in length. Witnesses report seeing its hard, jointed plates and numerous legs moving in a synchronized wave-like motion. Its color is often said to be dark, blending with the ocean depths, and some accounts mention spines or barbs along its body. Known for its intimidating size and unusual appearance, Con Rit is thought to be a lurking predator that may pose a threat to fishermen and seafarers.

Origin

The legend of Con Rit originates in the coastal regions of Vietnam and has circulated among fishermen and coastal communities for generations. Stories of this giant sea creature have persisted, with sightings occasionally reported throughout history. The first documented account of Con Rit in Western records dates back to the 19th century, when a French scientist wrote about a strange, elongated carcass that washed up on a Vietnamese beach, believed to be remnants of Con Rit. While no scientific evidence exists to confirm the creature's existence, some researchers speculate that the Con Rit legend could be based on sightings of oarfish or other large marine creatures with elongated, ribbon-like bodies similar to those described in Con Rit sightings.

Lore

Con Rit is often regarded as a powerful and fearsome figure in Vietnamese folklore, with stories emphasizing its immense strength and predatory nature. In some tales, the creature is seen as a guardian of the ocean, protecting the waters from those who disrespect the sea or harm marine life. Other accounts portray it as a danger to ships, using its powerful body to capsize boats or drag fishermen into the depths. Con Rit is usually described as elusive, appearing only during certain times or under specific weather conditions. Some modern interpretations have even likened Con Rit to a prehistoric creature, suggesting that it could be a survivor of ancient ocean life that has remained hidden in the deep seas.

Additional Facts
- Con Rit sightings have been reported in various coastal regions of Southeast Asia, particularly in Vietnam, where it holds a prominent place in local marine folklore.
- Similar cryptids exist in other cultures, such as the Akkorokamui of Japanese folklore, which is also described as a massive sea creature capable of capsizing boats.

Additional Resources

1. The book Cryptozoology: Sea Serpents and Lake Monsters by Gary J. Bradley explores marine cryptids, including Con Rit and its possible origins.

22

CORONADO SPHINX

(California, USA)

Description

The Coronado Sphinx is a mysterious cryptid reportedly seen along the beaches and coastal areas of Coronado, California. Described as a creature with a lion-like body and a human-like face, the Coronado Sphinx is often depicted with tawny fur, a muscular frame, and large, expressive eyes. Its face is said to possess human features, with an intense, piercing gaze and, in some accounts, a faint, unsettling smile. Although its exact size is debated, witnesses report it to be about the size of a large dog or mountain lion. Known for its elusive behavior, the creature typically appears at dusk or during foggy mornings, adding an air of mystery to its presence.

Origin

The legend of the Coronado Sphinx dates back to the early 20th century, with local rumors and sporadic sightings reported by residents and tourists along the southern California coast. While no verifiable origins explain its existence, some speculate that the creature may be a modern myth inspired by Coronado's famous Egyptian Revival architecture, which includes structures adorned with sphinx-like sculptures. Others believe the creature may be rooted in Native American lore, as early Indigenous groups often spoke of powerful animal spirits along the coastline. Reports of a "sphinx-like" being on the beach have persisted into recent years, fueled by tales from fishermen and beachgoers who claim to have encountered the mysterious creature.

Lore

The Coronado Sphinx is often seen as a guardian of the coastal lands, appearing to those who are lost or in need of guidance. Some locals claim that the creature appears in moments of danger, serving as a protective spirit. In other stories, the Coronado Sphinx is seen as a playful, if enigmatic, figure, vanishing into the mist if approached too closely. Over time, the legend has evolved, and the Coronado Sphinx is now part of local folklore, captivating the imaginations of residents and visitors alike. Sightings are said to occur more frequently during the winter months, when the coastline is quieter and the fog more frequent. Many believe that the Coronado Sphinx only reveals itself to those who truly respect the land, adding a sense of reverence to its legend.

Additional Facts
- Sightings of the Coronado Sphinx are typically reported near Coronado Beach, especially around the foggy early mornings and late evenings.

- Some locals attribute sightings to atmospheric phenomena or optical illusions, especially during foggy weather when shapes and shadows can appear distorted.

Additional Resources

1. Myths and Legends of California by Katharine Berry Judson explores regional folklore, including stories from California's Indigenous cultures that have inspired local cryptid lore.

23

CRESSIE

(Canada)

Description

Cressie, also known as the "Pond Monster," is a cryptid said to inhabit Crescent Lake in Newfoundland and Labrador, Canada. Described as a large, eel-like creature, Cressie is often portrayed with a long, sinuous body, dark scales, and a large head with a wide mouth. Sightings of Cressie have reported lengths of up to 15 feet or more, with some describing the creature as having a serpentine, undulating motion as it moves through the water. Its mysterious appearance and elusive behavior have sparked curiosity and fear among locals and visitors alike, making Cressie one of Canada's more intriguing lake monsters.

Origin

The legend of Cressie dates back to Indigenous Mi'kmaq and Beothuk stories, where it was seen as a guardian spirit of the lake or an ominous presence within its depths. These early tales spoke of a creature that would occasionally surface, instilling caution in those who ventured too close to Crescent Lake's waters. The name "Cressie" was later popularized by European settlers in the 19th and 20th centuries, who recorded their own encounters and sightings. Reports of Cressie continue to this day, with stories of the creature resurfacing every few years, keeping the legend alive in Newfoundland folklore.

Lore

Cressie is often considered both a protector of the lake and a foreboding figure, with local lore warning fishermen and swimmers to be cautious around Crescent Lake. Some tales suggest that Cressie only appears to those who disrespect the lake or attempt to harm its ecosystem, serving as a guardian of nature. Other stories describe it as a shy, reclusive creature that avoids human contact, surfacing only occasionally. In recent years, Cressie has gained popularity as a local mascot of sorts, with many viewing the creature as a unique symbol of Newfoundland's mysterious and untamed wilderness. Despite skepticism, sightings continue, and Cressie remains an enduring figure in local culture and storytelling.

Additional Facts

- Crescent Lake, home to Cressie, is located near Robert's Arm in Newfoundland, where many reported sightings occur.

- Cressie is sometimes compared to other lake monsters, such as Scotland's Loch Ness Monster and British Columbia's Ogopogo, though it is generally depicted as more eel-like than serpentine.

- The creature has inspired local artwork and folklore, appearing in stories, songs, and even tourism promotions, drawing visitors to the area.

Additional Resources

1. Lake Monsters and Mysteries by Benjamin Radford provides an analysis of lake monster legends, including Cressie.

2. The Newfoundland and Labrador Heritage Website offers information on Indigenous legends and folklore.

24

DEVIL MONKEY

(North America)

Description

The Devil Monkey is a cryptid reported throughout North America, particularly in rural areas of the United States. Described as a large, aggressive, and ape-like creature, the Devil Monkey is known for its dog-like face, large fangs, and pointed ears. Reports describe it as standing around three to five feet tall, with powerful hind legs similar to those of a kangaroo, allowing it to leap great distances. It is often said to have shaggy, dark fur and sharp claws. Witnesses frequently report a menacing demeanor, with some describing its glowing red eyes and loud, piercing screeches that add to its unsettling presence. Its aggressive behavior and unusual physical characteristics have made the Devil Monkey an enduring figure in American folklore.

Origin

The first documented sighting of a Devil Monkey occurred in 1934 in South Pittsburgh, Tennessee, when a local reported being attacked by a creature resembling a large monkey with a dog-like face. Sightings continued sporadically, with notable reports in the 1970s from Kentucky, where locals described encounters with a large, aggressive creature that could leap significant distances. In some regions, the Devil Monkey legend is thought to have been influenced by local folklore involving mysterious animals or spirits in the wilderness. Although sightings have decreased in recent years, reports continue to surface occasionally, primarily in Appalachian and Midwestern states.

Lore

In local lore, the Devil Monkey is often seen as a predatory creature, known to raid farms and terrorize livestock. Some tales suggest it has a taste for small animals and, occasionally, pets, leading to stories of missing animals in areas with reported Devil Monkey sightings. In regions where sightings are frequent, locals warn against venturing into the woods at night or leaving small animals unattended. Over time, the creature has gained a reputation as an aggressive and elusive beast, with its strange physical characteristics lending to theories that it might be a hybrid creature, an escaped exotic animal, or even a supernatural entity. Despite skepticism, stories of the Devil Monkey remain popular in local folklore and cryptid communities.

Additional Facts
- Sightings of the Devil Monkey have been reported in states across North America, including Kentucky, Tennessee, and Virginia, as well as parts of the Midwest.

- The creature's aggressive behavior and reported attacks on humans and animals have led locals to associate it with ominous or supernatural qualities.

Additional Resources

1. The Field Guide to North American Monsters by W. Haden Blackman includes accounts of the Devil Monkey and other regional cryptids.

25

DOBHAR-CHÚ

DOBHAR-CHÚ

Description

The Dobhar-Chú, sometimes referred to as the "Irish Water Hound" or "King Otter," is a cryptid from Irish folklore said to inhabit the lakes and rivers of Ireland. This mysterious creature is often described as a large, otter-like animal, with some accounts suggesting it is nearly as big as a human. The Dobhar-Chú is noted for its powerful, muscular build, sleek, dark fur, and a head that resembles a cross between a dog and an otter. Witnesses have reported sharp claws and prominent teeth, which, along with its speed and aggressive behavior, give the creature a fearsome reputation. Its most distinct feature is its alleged blood-curdling howl, said to echo across the water, warning of its presence.

Origin

The legend of the Dobhar-Chú dates back centuries, with some of the earliest references appearing in Irish oral tradition. The creature has long been feared by local communities, especially in rural areas where lakes and rivers play a central role in daily life. One famous account from the 17th century tells of a woman named Grace Connolly, who was reportedly killed by a Dobhar-Chú near Glenade Lake in County Leitrim. Her husband allegedly avenged her by tracking down and killing the beast, a tale commemorated on Grace's grave, where a carving of the creature can still be seen. This legend has contributed significantly to the Dobhar-Chú's enduring mystique in Irish folklore.

Lore

In Irish lore, the Dobhar-Chú is seen as a territorial and vengeful creature, attacking those who encroach upon its domain. Stories of the beast often portray it as a fiercely protective animal, especially if it perceives its mate or young to be threatened. Some tales describe the Dobhar-Chú traveling in pairs, with one companion seeking revenge if the other is killed. The creature is often likened to a spirit or guardian of Ireland's waterways, with locals believing that to encounter it is an omen of impending misfortune. Its association with tragedy and revenge has made it a formidable figure in Irish mythology, both feared and respected by those who live near its reputed habitats.

Additional Facts
- Sightings of the Dobhar-Chú are most often reported in rural Ireland, particularly in the lakes and rivers of counties Leitrim and Sligo.

- The creature's name, "Dobhar-Chú," roughly translates from Irish as "water hound," emphasizing its connection to aquatic environments.

Additional Resources

1. Irish Wonders by D.R. McAnally Jr. provides an overview of Irish myths and legends, including tales of the Dobhar-Chú and other water creatures.

26

DOGMAN

(Michigan, USA)

Description

The Dogman is a cryptid reported in Michigan folklore, described as a large, wolf-like creature with humanoid characteristics. Witnesses report it standing between six to seven feet tall when on its hind legs, with a muscular build, shaggy fur, and a canine face featuring a pronounced snout, sharp fangs, and glowing eyes. The Dogman is often seen moving both on two legs and all fours, with powerful limbs that allow it to run at high speeds. Its fur is typically dark, either black or dark brown, and it is known for an eerie growl or howl that some witnesses claim resembles a human scream. The creature's elusive and intimidating presence has led to a mix of fear and fascination among locals, making it one of Michigan's most well-known cryptids.

Origin

The legend of the Michigan Dogman dates back to Native American lore, which speaks of a werewolf-like creature that roamed the forests of the Great Lakes region. However, the modern Dogman story emerged in 1887 in Wexford County, Michigan, when two lumberjacks allegedly encountered a creature with a man's body and a dog's head. Since then, sightings have occasionally surfaced across Michigan, particularly in rural areas near dense forests. The legend gained significant attention in 1987 when radio DJ Steve Cook released a song, The Legend, based on Dogman encounters, which brought the cryptid into popular culture and sparked renewed interest in the creature's mystery.

Lore

Local lore portrays the Dogman as an elusive and often aggressive creature, one that guards the forests and occasionally terrifies those who wander too close to its domain. Tales from Michigan residents describe the creature as intelligent and highly aware of human presence, often watching from the shadows or even following people as they drive down isolated roads. In some accounts, the Dogman is seen as a supernatural entity or even a shape-shifter, linked to Native American beliefs about spirit guardians of the land. Its sporadic appearances, particularly in the northern part of Michigan, have led some to believe that it is tied to certain cycles or seasons, re-emerging under specific conditions.

Additional Facts
- The 1987 song The Legend by Steve Cook popularized the Dogman legend and led to an increase in reported sightings, many of which described the creature's menacing appearance.
- The Dogman's distinct scream or howl, which some say sounds disturbingly human, is a common feature in eyewitness accounts.

Additional Resources

1. The Michigan Dogman: Werewolves and Other Unknown Canines Across the U.S.A. by Linda S. Godfrey explores the Dogman and other canine cryptids in Michigan and beyond.

27

DOVER DEMON

(Massachusetts, USA)

Description

The Dover Demon is a cryptid known for its eerie, otherworldly appearance, reportedly seen in Dover, Massachusetts. Witnesses describe it as a small humanoid figure standing about three to four feet tall, with a thin, spindly body and long fingers. Its most striking features are its large, round head and enormous, glowing orange or green eyes, which dominate its face. The creature's skin is often described as smooth and hairless, with a pale or greyish tone, resembling rough sandpaper. Unlike most cryptids, the Dover Demon lacks a mouth, nose, or other facial features, further adding to its mysterious and unsettling appearance. Its strange, silent demeanor and limited sightings make it one of the more enigmatic figures in American cryptid lore.

Origin

The Dover Demon legend originated in 1977 when three teenagers reported separate sightings of the strange creature over two days in April. The first sighting was by 17-year-old Bill Bartlett, who spotted the creature perched on a stone wall while driving with friends. He described a thin figure with large, glowing eyes that stared directly at him. Later that night, John Baxter, another local teenager, reported seeing the creature standing near a wooded area. A third sighting followed, with Abby Brabham claiming to have encountered the same creature. These sightings were investigated by local authorities, but no evidence was found. The story quickly spread, drawing media attention and solidifying the Dover Demon as a notable American cryptid.

Lore

Though there were only a few sightings in 1977, the Dover Demon has since become a staple of cryptid lore in Massachusetts. The creature is often seen as an elusive, possibly extraterrestrial being, as its appearance and behavior align with descriptions of otherworldly entities. Some speculate that it could be an alien or an interdimensional traveler, visiting the area briefly before vanishing. Others view it as a lost or unknown species, explaining its strange features and silent behavior. In local folklore, the Dover Demon is considered a harmless yet unsettling presence, with no reports of aggression toward witnesses. Its limited appearances and lack of interaction with humans have kept its mystery alive, with many cryptid enthusiasts and paranormal investigators continuing to explore its story.

Additional Facts
- Sightings of the Dover Demon have remained limited to the 1977 incidents, with no other reports documented in recent years.
- The creature is often compared to other extraterrestrial or cryptid sightings due to its unusual, humanoid appearance and glowing eyes.
- The Dover Demon's unique appearance has inspired various depictions in pop culture, with its likeness appearing in artwork, books, and documentaries focused on American cryptids.

Additional Resources

1. The Field Guide to Extraterrestrials by Patrick Huyghe offers accounts of the Dover Demon and similar beings.

28

DROP BEAR(S

(Australia)

Description

The Drop Bear is a cryptid from Australian folklore, often portrayed as a larger, more aggressive version of the common koala. Described as a carnivorous marsupial with sharp claws, large teeth, and a muscular build, the Drop Bear is said to drop from trees onto unsuspecting prey. It is reportedly much larger than a typical koala, with some accounts describing it as nearly twice the size. Witnesses claim that it has coarse, dark fur and a menacing expression, giving it a fearsome reputation among locals and tourists alike. Known for its stealthy approach, the Drop Bear is said to remain motionless in trees until it spots a target, then drops down from above to ambush its prey.

Origin

The Drop Bear legend originated as a humorous and somewhat cautionary tale in Australian culture, believed to have been created by locals to playfully scare tourists. The stories are thought to date back to colonial times when settlers encountered Australia's unique and often intimidating wildlife. Over time, the tale of the Drop Bear became a popular piece of folklore, particularly among Australians who enjoy sharing it with newcomers. The Drop Bear legend has continued to evolve, with many locals embellishing stories to increase their scare factor. Though not a real creature, the Drop Bear has become an iconic part of Australian culture and is often referenced in tourism as a tongue-in-cheek warning to travelers.

Lore

In Australian lore, the Drop Bear is often presented as a predatory animal that specifically targets tourists and those unfamiliar with the land. Locals sometimes claim that Drop Bears are attracted to the smell of certain products or types of food, warning visitors to avoid scented products or to place utensils in their hair as a "defense." Some stories suggest the Drop Bear is a territorial animal that defends its domain by frightening off intruders. While Australians generally treat the Drop Bear legend with humor, the creature has become a cultural symbol representing Australia's wild and unpredictable nature. Many tourists are introduced to the Drop Bear as a rite of passage, with locals using it to add an element of mystery and adventure to the country's unique landscape.

Additional Facts
- The Drop Bear is often compared to the Yowie, another mythical Australian creature, though the Yowie is depicted as more of a cryptid while the Drop Bear is purely humorous folklore.
- Although fictional, the Drop Bear remains one of the most enduring tales in Australian culture, showcasing the country's dry sense of humor.

Additional Resources

1. The Australian Museum offers an online page on Drop Bears, humorously detailing the "creature's" habits and characteristics.

2. The book Drop Bears and Other Animals by Geoffrey Goss provides insight into Australian folklore and cryptids.

29

DUENDE

(Latin America)

Description

The Duende is a legendary figure in Latin American folklore, often described as a small, elf-like creature with mischievous tendencies. These tiny beings are typically said to be about one to two feet tall, with pointed ears, a wide mouth, and large, expressive eyes. Their appearance varies depending on the region, with some accounts depicting them as elderly with long beards, while others describe them as more childlike in appearance. Known for their love of pranks and tricks, Duendes are often seen wearing rustic, earth-toned clothing or, in some accounts, no clothes at all. They are said to inhabit rural areas, especially forests, caves, or abandoned homes, where they emerge at night to play tricks on unsuspecting humans.

Origin

The origins of the Duende legend trace back to the Iberian Peninsula, particularly Spain and Portugal, where similar creatures known as "Duende" were part of folklore long before European settlers brought these tales to Latin America. As the stories merged with Indigenous beliefs in Latin America, the legend of the Duende evolved to fit local landscapes and cultural nuances. In various Latin American countries, Duendes are seen as both playful and protective spirits, closely connected to the natural world. Their tales often serve as cautionary stories for children, teaching them to respect nature and avoid wandering into unknown areas alone.

Lore

In Latin American folklore, Duendes are known for their trickster nature, sometimes playing innocent pranks, such as hiding small items or creating eerie noises to startle people. However, some legends depict them as having darker motives, luring children into the woods or causing people to become lost. Many believe that Duendes can be both benevolent and mischievous, protecting those who respect nature while causing trouble for those who harm the environment. In some areas, Duendes are said to be guardians of lost treasures or sacred places. Parents often warn their children about Duendes to keep them from wandering too far into the wilderness or straying near dangerous places. Despite their trickery, Duendes are generally respected and treated with caution in Latin American culture, as offending them could invite trouble or misfortune.

Additional Facts
- In Mexico, Duendes are said to live in houses, often hiding in attics or behind walls, where they may knock or make noises to remind residents of their presence.
- In Central American cultures, Duendes are sometimes considered protectors of the forest, aiding those who respect the environment and frightening those who disrespect it.

Additional Resources

1. Latin American Folktales by John Bierhorst offers a collection of stories, including tales of Duendes and other supernatural beings in Latin American culture.

30

EYEWITNESS OF GUADALCANAL

(Solomon Islands)

Description

The Eyewitness of Guadalcanal is a cryptid from the Solomon Islands, specifically Guadalcanal, where local lore and rare reports describe it as a large, mysterious, ape-like creature. Known to locals as the "Gigantu" or "Giants of Guadalcanal," this being is often said to stand between eight and ten feet tall, with a muscular build and covered in thick, dark hair. Its features are often described as a mixture of human and ape, with a broad face, deep-set eyes, and a sloping forehead. Witnesses sometimes mention the creature's powerful limbs and large hands, which seem well adapted to its mountainous jungle habitat. Eyewitnesses report that the creature moves with surprising grace, blending into the dense foliage as if it is part of the forest itself.

Origin

Stories of the Eyewitness of Guadalcanal have been passed down through generations within the indigenous communities of the Solomon Islands. The creature's legend is deeply rooted in the local culture and spirituality, with some tribes viewing it as a guardian of the forest, while others see it as a potential threat. Although reports of encounters by Westerners are rare, interest in the creature's existence grew after World War II, when Allied soldiers stationed on the islands during the Battle of Guadalcanal reportedly experienced strange occurrences and glimpsed large, shadowy figures in the forests. The island's rugged terrain and dense rainforests contribute to the mystery, as much of the region remains remote and difficult to access.

Lore

In Solomon Island folklore, the Eyewitness of Guadalcanal is often regarded as a spiritual protector of the forest, deterring those who seek to exploit its resources. Some local tales describe it as highly intelligent, with a strong sense of territory and an ability to evade humans with ease. It is said to live in the mountainous regions and only comes closer to villages on rare occasions, often during certain times of the year or as a warning to those who threaten its domain. Villagers tell stories of eerie, guttural calls echoing through the mountains at night, and some claim that the creature possesses supernatural qualities, such as the ability to communicate telepathically with select individuals. While generally shy and reclusive, the Eyewitness of Guadalcanal is respected and sometimes feared, with locals believing that encountering it is a powerful and potentially life-altering experience.

Additional Facts
- Local stories suggest the creature has a complex social structure, with some villagers claiming to have seen smaller, younger versions of the cryptid, suggesting family groups.

- Reports of strange footprints, broken tree branches, and eerie noises attributed to the creature have been documented by explorers and locals, though photographic evidence remains elusive.

Additional Resources

1. The book The Solomon Islands Mysteries by Marius Boirayon provides an in-depth look at the cryptid stories.

31

FAIRY

(Europe)

Description

Fairies, often depicted as small, ethereal beings with delicate wings, are cryptids rooted in European folklore but have become legendary worldwide. They are typically described as miniature human-like creatures, adorned in natural attire made from leaves, flowers, or other elements of the forest. Depending on the culture and story, fairies vary in appearance and size, with some being as small as an insect and others standing several feet tall. Known for their radiant beauty, sparkling aura, and gossamer wings that allow them to fly, fairies are said to possess magical abilities, such as invisibility, shape-shifting, and the power to influence the human world. While often associated with kindness and whimsy, some stories paint them as mischievous or even malicious tricksters, particularly if they are disrespected or their territory is disturbed.

Origin

The legend of fairies dates back to ancient Europe, particularly in Celtic and Norse mythologies, where they were believed to inhabit the "Otherworld" or magical realms hidden within nature. In Ireland and Scotland, they are considered part of the Aos Sí, a supernatural race closely linked to the ancestors and natural world, while Norse mythology describes similar beings like the Alfar or Elves. The Victorian era in Britain saw a resurgence of interest in fairies, romanticizing them as gentle, harmless creatures, though older tales portrayed them as powerful, unpredictable spirits. The fairy mythos has since spread globally, evolving through local folklore, literature, and popular media to include a wide range of variations, from benevolent nature spirits to mischievous creatures known for playing tricks on humans.

Lore

In fairy lore, these beings are often described as guardians of nature, residing in hidden, sacred places like ancient forests, groves, or secluded glens. They are said to live in communities and celebrate festivals, following their own customs and hierarchy. Fairies are known for their trickster nature and can be both helpful and harmful, depending on how they are treated by humans. They are believed to reward those who show respect for nature and punish those who harm their surroundings. In many traditions, fairies are known for "borrowing" small items from humans, like shiny objects, or leading travelers astray with dancing lights.

Additional Facts

- Fairy rings, or naturally occurring circles of mushrooms, are often associated with fairies and are considered magical. Folklore suggests that humans should avoid stepping into these rings, as they are believed to be portals to the fairy world.

- Offerings of milk, honey, or small trinkets are sometimes left out to appease fairies or gain their favor, particularly in rural areas where fairy legends are prevalent.

Additional Resources

1. The book Fairy and Folk Tales of the Irish Peasantry by W.B. Yeats compiles traditional fairy tales and folklore, providing insight into Irish fairy beliefs.

32

FANG SNAKE

(CHINA)

Description

The Fang Snake is a cryptid from Chinese folklore, often described as a massive serpent with formidable fangs and an aggressive demeanor. Typically, this creature is depicted as being much larger than typical snakes, with some accounts claiming it can reach lengths of 20 feet or more. The Fang Snake is said to have a sleek, dark body with scales that gleam ominously in the light, and its fangs are reported to be long and razor-sharp, capable of delivering a deadly bite. Witnesses often mention its piercing yellow or green eyes that glow in low light, adding to its fearful reputation. Known for its stealth and speed, the Fang Snake is believed to be highly venomous and prone to ambush, lurking in secluded forests, caves, or mountainous regions.

Origin

The Fang Snake legend has roots in ancient Chinese folklore, where large serpents have often been depicted as powerful, sometimes supernatural beings. This creature has long been associated with the mountainous and forested regions of China, especially near remote villages where it is said to roam. Local legends tell of a fearsome snake spirit that guards its territory and only reveals itself when threatened or provoked. Stories of the Fang Snake were likely influenced by actual encounters with large snakes, with the myth evolving over time to incorporate supernatural aspects. The creature's reputation spread through word of mouth and ancient texts, becoming a symbol of danger in the wilderness.

Lore

In Chinese folklore, the Fang Snake is seen as both a natural predator and a guardian of its territory, with some stories likening it to a spirit that protects the mountain forests. It is believed that those who encounter the Fang Snake and escape unscathed are blessed with luck and strength, though others warn that any confrontation could be deadly. Some villagers leave offerings to the creature to appease it and ensure safe passage through its domain. The Fang Snake is sometimes thought to possess mystical qualities, such as the ability to control its poison, using it either to heal or harm. Stories caution travelers to stay vigilant and respectful of the creature's territory, as disturbing its habitat could provoke its wrath.

Additional Facts
- The Fang Snake is sometimes compared to the Mangshan pit viper, a venomous snake found in southern China, which might have inspired or contributed to the legend.
- In some myths, the Fang Snake is said to have a spiritual connection to dragons, symbolizing both the fearsome and revered aspects of nature.

Additional Resources

1. Chinese Myths and Legends by Shelley Fu offers a detailed look at traditional Chinese folklore, including serpentine creatures like the Fang Snake.

33

FLATWOODS MONSTER

(West Virginia, USA)

Description

The Flatwoods Monster, also known as the Braxton County Monster, is a cryptid and alleged extraterrestrial being that gained notoriety after a sighting in Flatwoods, West Virginia, in 1952. Described as a tall, humanoid figure standing about 10 feet tall, the creature was said to have a distinct spade-shaped or hood-like structure around its head and glowing red eyes. Witnesses reported that the monster appeared to hover or float above the ground, emitting an eerie greenish glow. Its body was reportedly metallic or armored, resembling a suit or exoskeleton, with claw-like hands and a dark green or black color. A pungent, sulfurous odor was also reported at the sighting site, adding an unsettling detail to the encounter.

Origin

The Flatwoods Monster legend began on the evening of September 12, 1952, when several local boys and a mother in Flatwoods saw a bright light streak across the sky and land on a nearby hill. Curious, the group ventured to the site, where they reportedly encountered the monstrous figure. Frightened, they fled and reported the sighting to the authorities, which led to an investigation. Although officials dismissed it as likely a misidentified owl or other wildlife, the encounter captured national attention, with newspapers dubbing the creature the "Flatwoods Monster." Over the years, the story has become part of West Virginia folklore, with some speculating that it was an extraterrestrial being or a government experiment gone awry.

Lore

The Flatwoods Monster is often associated with UFO lore, and many believe it to be an extraterrestrial being rather than a cryptid. The initial encounter, along with subsequent UFO sightings in the area, contributed to a growing legend that the creature was an alien visitor. In local folklore, the monster is seen as both terrifying and awe-inspiring, an enigma that appeared once and then vanished. Some locals speculate that the creature may have arrived as part of an alien exploration or crash landing, while others believe it was a warning or symbol of things beyond human understanding. Though the creature has never been reported again, the story has inspired curiosity and fear, drawing enthusiasts and tourists to the small town of Flatwoods.

Additional Facts
- The monster's sulfurous odor has led to theories that it could be related to extraterrestrial or experimental gas leaks, though no evidence was ever found to support this.

- Despite skepticism, the encounter remains one of the most famous UFO and cryptid cases in the U.S., often cited alongside the Mothman of West Virginia.

Additional Resources

1. The Flatwoods Monster: A Legacy of Fear by George Dudding offers an in-depth look at the legend, investigating the 1952 sighting and subsequent theories.

34

FRESNO NIGHTCRAWLER

(California, USA)

Description

The Fresno Nightcrawler is a cryptid from California, often described as a tall, thin, humanoid figure with unusually long legs and a small upper body. Its most distinctive feature is its lack of a discernible head, with many descriptions suggesting it is simply a pair of elongated legs that taper off at the top, creating an almost ghostly silhouette. Typically white or pale gray in color, the creature is said to move with a smooth, gliding motion, making it appear as though it is floating or walking in an exaggerated, stilt-like manner. Sightings of the Nightcrawler are rare and mostly limited to security footage, adding an eerie, surreal quality to its legend.

Origin

The legend of the Fresno Nightcrawler began in the late 2000s, with the first known footage surfacing in Fresno, California. A homeowner's security camera reportedly captured two strange, leg-like figures walking across their yard. This footage quickly gained attention on the internet, sparking discussions and theories about the creatures' origins. Over time, similar sightings and video recordings have emerged, mainly in California, though some reports suggest appearances in other parts of the United States. The cryptid's sudden emergence and lack of historical accounts add to its mystery, making it a modern yet enigmatic figure in American cryptid lore.

Lore

In local lore, the Fresno Nightcrawler is seen as a peaceful, otherworldly entity. Some believe it is an alien or interdimensional being, observing our world without interacting. The creature's smooth, rhythmic movement has led to theories that it is a spiritual being, more connected to nature than to humans. Unlike many cryptids that are associated with danger or malice, the Fresno Nightcrawler is often perceived as benign, almost curious in its appearance. Some locals speculate it has a purpose yet unknown, with some folklore suggesting it could be a guardian of the natural world. Its unique, nonthreatening presence has contributed to its intrigue, inspiring curiosity rather than fear.

Additional Facts
- The creature's unusual gait and appearance have led some to suggest it could be a misidentified animal, a puppet, or a hoax, though no clear explanation has been confirmed.
- Similar sightings have been reported as far away as Yosemite National Park, though these accounts remain unverified.
- Cryptozoologists and paranormal enthusiasts continue to investigate the Nightcrawler, with some theorizing that it may belong to an unknown species of life form.

Additional Resources
1. The book Monsters of California by Peter Byrne covers regional cryptids, including a section on the Fresno Nightcrawler and its cultural impact.

35

GLOUCESTER SEA SERPENT

(Massachusetts, USA)

Description

The Gloucester Sea Serpent is a legendary cryptid reported off the coast of Gloucester, Massachusetts, with sightings dating back to the early 19th century. Witnesses describe it as a massive, snake-like creature with a long, undulating body that can measure up to 100 feet in length. Accounts mention dark, scaly skin, a head resembling that of a horse or snake, and large, bulging eyes. Some sightings report a series of humps visible above the waterline, giving it a serpentine appearance as it glides through the waves. The creature is said to move swiftly and gracefully, with a distinctive "rolling" motion across the water's surface.

Origin

The legend of the Gloucester Sea Serpent began in 1817, when numerous sightings were reported by residents of Cape Ann. Locals, including fishermen and other seafarers, claimed to have seen the creature swimming near the shore. The most notable encounter occurred in August of that year, when a group of citizens observed the creature from a beach, noting its large, snake-like form. Following these sightings, newspapers across New England covered the story, sparking widespread interest and speculation. The Linnaean Society of New England investigated the reports but found no concrete evidence. Despite this, the sightings became an enduring part of Massachusetts folklore.

Lore

The Gloucester Sea Serpent is often regarded as a mysterious guardian of the Massachusetts coast, tied to the untamed and wild aspects of the sea. Some stories suggest it may be a relic of ancient marine life, surfacing only rarely. Others view it as a symbol of the unknown dangers lurking beneath the ocean, a reminder of the mysteries that remain hidden in the world's waters. For locals, the sea serpent legend has become part of Gloucester's maritime heritage, capturing the curiosity and fear of residents and visitors alike. Although sightings have declined over time, occasional reports continue to keep its intrigue alive.

Additional Facts

- The Linnaean Society's investigation in 1817 resulted in the tentative classification of a mysterious "Scoliophis atlanticus," later dismissed due to insufficient evidence.

- The legend has inspired local art and maritime memorabilia in Gloucester, celebrated in local folklore and occasional festivals.

- The Gloucester Sea Serpent is often compared to other legendary sea cryptids, such as the Loch Ness Monster, due to its size, shape, and elusive nature.

Additional Resources

1. Monsters of the Sea by Richard Ellis explores the history of sea serpent sightings, including a section on the Gloucester Sea Serpent.

2. The Cape Ann Museum in Gloucester has exhibits on local maritime history, including the Gloucester Sea Serpent.

36

GOBLIN

(Europe)

Description

The goblin is a cryptid known across Europe, often described as a small, elusive creature with a mischievous or even malevolent disposition. Unlike familiar folklore depictions, which often portray goblins as impish troublemakers, cryptid sightings depict goblins as more physical and unpredictable beings. Witnesses commonly describe them as squat, humanoid figures around three feet tall, with leathery, often wrinkled skin, sharp features, and large, glinting eyes adapted for low light. Goblins are often said to have spindly limbs, claw-like fingers, and sharp teeth that add to their eerie, sometimes menacing, appearance. Their skin color varies from dark green to earthy brown, blending seamlessly into shadowy environments. They are reputed to move with agility, making them difficult to spot clearly and contributing to their mysterious reputation.

Origin

Goblin legends date back to early medieval Europe, with references in folklore, literature, and oral traditions from countries like France, Germany, and the British Isles. While folklore often portrays goblins as spirits or magical beings, cryptid sightings have given rise to the theory that goblins might be real, undiscovered creatures rather than mythical ones. Reports of goblins appearing in forests, caves, and abandoned buildings have circulated for centuries, especially in rural areas where strange sightings are less likely to be dismissed as imagination. Over time, tales of these creatures evolved, mixing folklore with possible sightings, creating a blend of mystery and fear surrounding the goblin cryptid.

Lore

In cryptid lore, goblins are regarded as reclusive yet cunning creatures, known to inhabit secluded or dark areas such as caves, dense forests, and even the basements of old houses. They are often associated with pranks and disturbances, such as items going missing or strange noises echoing through the night. Some believe that goblins are highly territorial, protecting their claimed areas by frightening away intruders with bizarre sounds or fleeting shadows. While generally viewed as solitary, goblins are sometimes rumored to live in small groups, working together to defend their lairs.

Additional Facts
- Some researchers speculate that goblin sightings may be misidentifications of nocturnal animals or even undocumented species adapted to low-light environments.
- Goblins have been associated with other cryptids, such as trolls or fairies, due to their size and elusive nature, though they are often distinguished by their sharper, more humanoid features.

Additional Resources

1. European Cryptids and Legends by Martin J. Tyler explores the history and sightings of lesser-known cryptids, including goblins and other mysterious beings.

37

GROOTSLANG

(South Africa)

Description

The Grootslang is a cryptid from South African folklore, often described as a massive, snake-like creature with elements resembling an elephant or dragon. Legends say it can reach lengths of up to 40 feet or more, with a thick, scaly body and large, powerful jaws filled with sharp teeth. Some accounts claim it has tusks like an elephant, adding to its terrifying appearance. The Grootslang is typically reported to inhabit caves and deep, secluded areas, particularly around the Richtersveld region, where its lair, known as the "Bottomless Pit," is said to reside. Known for its cunning and ferocity, the creature is feared for its potential to ambush anyone who strays too close to its domain.

Origin

The Grootslang legend is believed to stem from ancient myths among the Khoikhoi and San people of South Africa. According to legend, the Grootslang was one of the first creatures created by the gods, who gave it immense strength and intelligence. Realizing their mistake, the gods split the creature into two separate species—snakes and elephants. However, one Grootslang escaped the gods' intervention and continued to roam the depths of South Africa. In the early 20th century, stories of the Grootslang resurfaced, capturing the attention of European explorers and treasure hunters intrigued by tales of a monster guarding a diamond-filled cave. The creature has since become an enduring part of South African folklore.

Lore

The Grootslang is often seen as a protector of its territory, fiercely guarding its lair and the wealth it is rumored to hide. Legends say that it hoards precious gems, particularly diamonds, within its cave, drawing adventurers despite the dangers. Some stories claim the creature can be appeased by offering precious stones, allowing a rare few to escape unharmed. In local lore, the Grootslang is both feared and respected, symbolizing the untamed power of nature and the mystery of the unexplored. Although its presence is tied to certain caves in South Africa, sightings and stories of the Grootslang have fueled interest in cryptid circles and inspired expeditions to locate its legendary lair.

Additional Facts
- The Grootslang is sometimes compared to other legendary serpentine creatures worldwide, though it is unique in its fusion of elephantine and serpentine traits.
- In South African folklore, some believe the Grootslang has mystical properties, allowing it to disappear into the depths of the earth or swim vast distances underground.

Additional Resources

1. African Myths and Legends by Kathleen Arnott explores traditional African folklore, including cryptids.

38

HELLHOUND

(Europe)

Description

The Hellhound is a fearsome cryptid from European folklore, often depicted as a large, spectral dog with a dark or black coat, glowing red or green eyes, and an otherworldly presence. It is commonly described as having a powerful, muscular build, capable of moving swiftly and silently, and sometimes leaving scorch marks or faint trails of smoke in its wake. Witnesses report a pungent sulfur-like odor accompanying the creature, as well as an intense feeling of dread when it appears. Hellhounds are frequently associated with graveyards, crossroads, or remote locations, where they are believed to roam as ominous guardians or harbingers of doom.

Origin

The legend of the Hellhound has roots in ancient European folklore, where it has long been regarded as a supernatural being rather than a mere animal. Stories of ghostly or infernal dogs are particularly prevalent in Britain, where they are known by various names such as Black Shuck, Barghest, and Padfoot. In Norse mythology, Hellhounds are thought to be related to Garmr, the guardian of Helheim, the realm of the dead. Many of these tales portray the Hellhound as a creature connected to the afterlife, either as a servant of death or as a guardian to ward off evil spirits. Over time, the Hellhound legend spread across Europe, with variations in different regions, though its ominous associations with death and the supernatural remain consistent.

Lore

In European folklore, the Hellhound is often seen as a creature to be avoided at all costs, with sightings considered an ill omen or even a warning of impending death. In some regions, the Hellhound is believed to be the guardian of sacred or cursed places, protecting the dead or guarding treasures buried in ancient graves. Encounters with Hellhounds are generally brief, and the creature is known to vanish without a trace, leaving only scorched earth or claw marks as evidence of its presence. Tales of the Hellhound frequently warn against provoking the creature, as it is believed to have supernatural strength and the ability to vanish into thin air. Though typically solitary, some stories describe packs of Hellhounds patrolling desolate areas or appearing during storms and other ominous weather conditions.

Additional Facts

- Hellhound legends are particularly strong in the United Kingdom, with local names such as Black Shuck in East Anglia and the Moddey Dhoo on the Isle of Man.
- In German folklore, a similar spectral dog known as the Schwarzer Hund is said to guard graveyards, further linking the Hellhound to the realm of the dead.

Additional Resources

1. The Black Dog Folklore of the British Isles by Mark Norman explores Hellhound myths and their cultural impact across the UK.

39

HODAG

(Wisconsin, USA)

Description

The Hodag is a legendary creature and cryptid from Rhinelander, Wisconsin, known for its bizarre and fearsome appearance. It is described as a beast with a reptilian body, spikes along its back, and the head of a bull or frog with prominent horns. The Hodag reportedly has razor-sharp claws, a long tail, and dark, leathery skin. Some descriptions include glowing red eyes and a pungent sulfurous smell, adding to its fearsome reputation. It is said to be about the size of a large dog or small bear, known for its growls and other unsettling noises echoing through the woods.

Origin

The Hodag legend originated in the late 19th century, when Wisconsin land surveyor and timber cruiser Eugene Shepard claimed to have discovered the creature. In 1893, Shepard presented the Hodag to the public with an elaborate hoax, complete with a staged "capture" and photographic evidence. This quickly became a local sensation, drawing curious visitors to Rhinelander. Although later revealed as a fabrication, Shepard's story captured the imagination of the town, and the Hodag became a cherished part of local folklore. Despite its dubious origins, the creature remains an iconic figure in Wisconsin's culture and tourism.

Lore

The Hodag is typically portrayed as a reclusive and territorial creature, lurking in the dense forests of the northern Midwest. Tales describe it as having an appetite for mischief and a fondness for terrorizing local loggers, though it is not considered a true threat. Many believe the Hodag represents the spirit of Wisconsin's wilderness, embodying the mystery and ruggedness of the region's deep woods. Although initially a hoax, the creature has become a symbol of local pride and folklore, with many legends, stories, and sightings woven around it. The Hodag's presence is celebrated in Rhinelander, with festivals, statues, and merchandise dedicated to this unique piece of local mythology.

Additional Facts

- Sightings of the Hodag are largely regarded as local lore, and it is generally accepted that the creature's origins are a playful piece of Wisconsin history.

- The Hodag's legend has made Rhinelander famous, with a large statue of the creature and an annual Hodag Country Festival drawing tourists each year.

- In modern culture, the Hodag is frequently associated with other North American cryptids, like Bigfoot or the Jersey Devil, though its story has a distinct blend of humor and tall-tale spirit.

Additional Resources

1. The book Fearsome Critters by Henry Tryon includes an entry on the Hodag and similar North American creatures.

40

IGOPOGO

(Canada)

Description

The Igopogo, also known as the "Kempenfelt Kelly" or "Beaverton Bessie," is a lake monster reputed to dwell in the depths of Lake Simcoe, Ontario, Canada. Described as an aquatic creature with a slender, snake-like body, Igopogo is often reported to have a dog-like or horse-shaped head with large eyes and sometimes a snout or whiskers. Its body is said to measure between 12 to 40 feet in length, covered in smooth, dark skin, with humps visible above the water when it surfaces. Some witnesses claim it has flippers or fin-like appendages, while others describe it as more eel-like. Igopogo has earned a reputation as a shy creature, rarely seen near populated areas, and is often spotted on foggy mornings or late at night.

Origin

The Igopogo legend dates back to Indigenous oral traditions, which speak of mysterious creatures inhabiting Lake Simcoe and other Canadian lakes. Local folklore surrounding this lake monster gained more traction during the 19th century when settlers began reporting sightings. Interest in Igopogo continued to grow throughout the 20th century, with numerous accounts from fishermen and boaters claiming to have seen a strange creature swimming in the lake. The name "Igopogo" emerged as a playful nod to other North American lake monsters like Ogopogo and Bigfoot, cementing its place as part of Canadian cryptid folklore.

Lore

In Canadian folklore, Igopogo is seen as a harmless but elusive creature, rarely venturing close to humans. Many locals regard it as a peaceful lake spirit, occasionally surfacing to surprise boaters or fishermen. Some stories describe Igopogo as a guardian of the lake, protecting its natural balance and warning those who might disrespect the waters. Over time, sightings of Igopogo have inspired a mix of fear, curiosity, and respect, with some suggesting that it represents the mysterious and ancient lifeforms hiding in Canada's northern lakes. Igopogo's enduring presence in local lore and occasional sightings keep its legend alive, drawing curiosity and fascination to Lake Simcoe.

Additional Facts
- The creature is often compared to other lake cryptids, such as the Loch Ness Monster and British Columbia's Ogopogo, due to its serpentine shape and mysterious nature.
- Lake Simcoe's reputation as the home of Igopogo has drawn cryptid enthusiasts, who occasionally conduct research or expeditions to seek evidence of the creature.

Additional Resources

1. Canadian Monsters and Mysteries by John Robert Colombo includes accounts of Igopogo and other legendary creatures in Canada.

2. The Lake Simcoe Region Conservation Authority occasionally shares folklore related to the lake.

41

INUIT GIANT

(Inuit Regions)

Description

The Inuit Giant, also known as the Tornit or Tuniit in Inuit mythology, is a legendary being known for its enormous size, strength, and mysterious nature. Stories describe the giants as towering figures, often two to three times the height of a human, with massive, muscular builds and rough, weathered skin suited to the harsh Arctic environment. Inuit Giants are sometimes portrayed as having long hair, fierce expressions, and wearing simple clothing made from animal skins. They are generally depicted as reclusive beings, living in remote, mountainous areas, caves, or the vast Arctic tundra. Despite their intimidating appearance, the giants are often described as solitary, preferring to avoid human contact.

Origin

The legend of the Inuit Giant has roots in ancient Inuit folklore and oral traditions, passed down through generations. According to legend, the Inuit Giants were an early race of people who lived in the Arctic regions before the arrival of the Inuit. Known as the Tornit or Tuniit, they are believed to have been mighty hunters and skilled craftsmen but ultimately retreated from human settlements, disappearing into the wilderness. Some believe they left due to conflict with the Inuit people, while others suggest they moved to escape the encroachment of modern life. Archaeological evidence of ancient tools, large footprints, and abandoned shelters in the Arctic has fueled speculation about the existence of a mysterious ancient people.

Lore

In Inuit lore, the giants are often portrayed as fierce yet tragic figures, connected to the land and possessing a unique bond with nature. Stories tell of their incredible strength and hunting skills, with some legends suggesting they could lift large boulders, travel vast distances swiftly, and endure the extreme Arctic cold. However, the giants are also said to have a certain sadness, often seen as lonely and displaced from their once-thriving society. Many tales depict them as wary of humans, hiding in secluded areas and rarely approaching villages. Inuit elders sometimes warn children not to wander too far, telling them that the giants may still roam the remote landscapes, watching over their former lands.

Additional Facts
- Inuit Giants are often associated with other Arctic cryptids or mythical beings, like the Qalupalik, due to their elusive nature and connection to the wild.
- The Tornit or Tuniit legends often serve as cautionary tales, symbolizing the struggle between tradition and change within Inuit culture.

Additional Resources

1. The Inuit Myth and Legend series by Rachel Qitsualik-Tinsley delves into Inuit folklore, covering legends of the Inuit Giants and other mythological beings.

42

JERSEY DEVIL

(New Jersey, USA)

Description

The Jersey Devil is a cryptid and legendary creature that has haunted the Pine Barrens of New Jersey for over 250 years. It is often described as a tall, kangaroo-like figure with the head of a horse, large, bat-like wings, horns, small arms with clawed hands, and cloven hooves. With glowing red eyes and a high-pitched, blood-curdling scream, the Jersey Devil's appearance and sounds strike terror in those who encounter it. The creature is said to move swiftly, both on the ground and in the air, and leaves behind strange hoof prints. The Jersey Devil is rumored to be highly elusive, appearing in wooded areas and secluded roads late at night.

Origin

The legend of the Jersey Devil dates back to the 1700s, with the most famous origin story involving a woman named Mother Leeds, who cursed her 13th child during childbirth, declaring it would be the "devil." According to the tale, the child was born as a monstrous creature and flew out of the house into the Pine Barrens, where it has roamed ever since. Other versions of the legend suggest the creature may be a supernatural being or even an early example of an American folk devil. The Jersey Devil gained national attention in 1909 when numerous sightings and encounters were reported across New Jersey and Pennsylvania, sparking fear and fascination throughout the region.

Lore

In local folklore, the Jersey Devil is often viewed as a harbinger of misfortune, appearing before disasters or during times of hardship. Many residents of the Pine Barrens have tales of strange encounters, eerie sounds, and unexplained events attributed to the creature. The Jersey Devil is also said to be highly territorial, protecting its wooded domain from intruders, and some even claim that it's responsible for livestock deaths and crop destruction in rural New Jersey. The creature has become a symbol of local pride and mystery, with annual festivals, tours, and even a minor league hockey team named in its honor. While sightings have decreased in recent years, the Jersey Devil remains a beloved figure in New Jersey folklore, inspiring numerous books, films, and paranormal investigations.

Additional Facts
- The Jersey Devil is known for its distinctive scream, often described as a mix of a woman's wail and an animal's shriek, echoing eerily through the woods.
- During the "Jersey Devil Panic" of 1909, hundreds of people reported sightings and experiences, leading to school and factory closures throughout the region.

Additional Resources

1. The Pine Barrens: Tales of the Jersey Devil and Other Legendary Creatures by John McPhee examines the history and myths surrounding the Jersey Devil and the Pine Barrens region.

2. The New Jersey Folklore Society offers resources and information on the Jersey Devil legend.

43

JUBOKKO

(Japan)

Description

The Jubokko is a vampiric tree spirit from Japanese folklore, known for its insidious ability to draw sustenance from human blood. Though it appears like an ordinary tree, it is said to have twisted branches and roots that can ensnare and drain those who come too close. Jubokko are often found on old battlefields or sites of great suffering, where they absorb the blood of the fallen. Its leaves are dark and glossy, sometimes said to glisten with a sinister red sheen, reflecting its bloodthirsty nature. When it attacks, the Jubokko uses its roots and branches to pierce the skin of its victim, sucking their blood like a vampire.

Origin

The legend of the Jubokko originates in ancient Japanese folklore, with stories emerging from regions marked by historic battles and natural disasters. The tree is said to have been ordinary before absorbing the spilled blood on battlefields, which transformed it into a cursed, predatory being. Over time, Jubokko tales became common cautionary stories, warning people against wandering alone in places where tragedy had struck. These legends reflect the spiritual beliefs in Japan about nature absorbing the essence of life and energy from events, both good and bad.

Lore

In Japanese lore, the Jubokko is feared for its seemingly innocuous appearance, as it is nearly impossible to distinguish from regular trees. Locals believe that it can live for centuries, only revealing its true nature when someone comes too close to its roots. Some stories suggest that the Jubokko is aware of its surroundings and that it even targets specific individuals—often those who disrespect the land. People claim that they can detect its presence by an inexplicable sense of dread or an eerie silence that falls in the area around it. While Jubokko are rarely defeated in traditional stories, it is said that fire or holy symbols can drive it back, restoring peace to the haunted site.

Additional Facts
- Some believe that the Jubokko is a type of yokai, or spirit, associated with vengeance, symbolizing nature's retribution for human violence.

- In some stories, people who approach the tree hear faint whispers or feel the ground pulsing, as if alive with the heartbeat of the trapped souls.

- The tree is sometimes said to produce sap that resembles blood, and local superstitions suggest avoiding such trees at all costs.

Additional Resources

1. The Yokai Museum: The Art of Japanese Supernatural Beings includes entries on the Jubokko and other mystical trees.

2. The book Japanese Ghosts and Demons by Stephen Addiss explores spirits like the Jubokko, and their origins.

44

JUMBO BIRDS

(Uganda)

Description

Jumbo Birds, as they are known locally, are legendary avian cryptids reported in various regions of Uganda. These mysterious birds are described as massive, with wingspans reportedly stretching up to 15 feet or more. They are often depicted with dark, glossy feathers, powerful talons, and piercing eyes that seem to glow at night. Sightings typically describe them as having broad, strong wings and long, sharp beaks, enabling them to hunt with precision. Some witnesses claim these birds resemble large eagles or vultures but are far larger, with an imposing presence that intimidates those who see them. Jumbo Birds are known to be nocturnal, swooping down from high perches and disappearing into dense forests or mountains.

Origin

The legend of the Jumbo Birds dates back centuries among the people of Uganda, where large, mysterious birds are often intertwined with folklore. Stories of enormous bird-like creatures have circulated in rural communities, especially among hunters and herders who occasionally report strange encounters in the wilderness. Local lore suggests that the Jumbo Birds are ancient protectors of the land, possibly linked to spiritual beliefs or ancestral spirits. In the 20th century, reports of large bird sightings intrigued researchers and travelers, though no confirmed evidence of these creatures was ever found. Some speculate that the legend might be based on sightings of the marabou stork, crowned eagle, or other large African birds, but none fully match the imposing size and mysterious nature of the Jumbo Birds.

Lore

In Ugandan lore, Jumbo Birds are often seen as guardians of the wilderness, revealing themselves only to those who venture too close to their territory. Some stories depict them as territorial and aggressive, defending remote regions of Uganda's forests and mountains. It is believed that the birds possess a supernatural awareness of their surroundings and can evade detection by humans. Local hunters and travelers often tell tales of hearing low, haunting cries at night, and some claim to have seen the dark silhouette of a giant bird against the moonlit sky. While primarily feared, Jumbo Birds are also respected, with some Ugandan traditions advising people to avoid their known territories and to treat sightings as a rare and powerful omen.

Additional Facts
- Local stories claim that Jumbo Birds can silently glide, adding to their eerie reputation as mysterious, almost ghostly beings.
- In certain communities, spotting a Jumbo Bird is considered a bad omen, potentially foretelling misfortune or warning of nearby dangers.

Additional Resources
1. The Ugandan Wildlife Authority occasionally publishes information on local folklore and legends, including cryptids and rare animal sightings in the area.

45

KASAI REX

(Congo)

Description

The Kasai Rex is a cryptid reported in the jungles of the Democratic Republic of the Congo, believed to resemble a living dinosaur. Described as a large, reptilian creature, the Kasai Rex is often said to resemble a Tyrannosaurus rex, standing on two powerful legs with a massive, muscular body. Its head is said to have a fearsome, crocodile-like appearance, with sharp teeth and powerful jaws capable of crushing large prey. Witnesses have described the Kasai Rex as covered in dark, scaly skin, with occasional sightings reporting reddish or brown patches. Known for its aggressive behavior, the creature is said to hunt and consume large animals such as antelope, leaving behind torn vegetation and tracks that have fueled speculation about its existence.

Origin

The first reported sighting of the Kasai Rex dates back to the 1930s, when Swedish hunter John Johnson allegedly encountered the creature near the Kasai Valley in the Congo. According to Johnson, the creature attacked and killed a rhinoceros, demonstrating its formidable strength and predatory instincts. While many skeptics regard Johnson's account as a tall tale or exaggeration, the story quickly became part of local lore and inspired further interest in the possibility of dinosaurs surviving in the dense jungles of Central Africa. Over the years, other reports of similar dinosaur-like creatures in the region have kept the legend alive, with some researchers speculating that the Kasai Rex could be an undiscovered species or a relic from a long-passed era.

Lore

In local Congolese lore, the Kasai Rex is often considered a powerful and dangerous spirit of the forest, a creature that serves as both a guardian and a warning to those who venture too deep into its territory. Some believe it to be a symbol of the wild, untamed nature of the Congo's jungles, a land where ancient secrets remain hidden from the modern world. Reports of the creature tend to emerge from remote, uncharted areas, adding to its mystique and the fear it inspires among locals. Though sightings are rare and unverified, the legend of the Kasai Rex has inspired numerous expeditions and cryptozoological inquiries, driven by the tantalizing idea of a dinosaur surviving in the modern age.

Additional Facts
- Skeptics argue that reports of the Kasai Rex could be attributed to misidentified large reptiles or even exaggerated encounters with known species such as monitor lizards or Nile crocodiles.

- The creature's ferocious reputation and alleged rhinoceros predation have contributed to its legendary status, setting it apart from other cryptids in the region.

Additional Resources

1. Living Dinosaurs? The Search for Prehistoric Relics by Roy Mackal offers an exploration of cryptids like the Kasai Rex, examining theories about surviving dinosaurs in remote areas.

46

KELPIE

(Scotland)

Description

The Kelpie is a shape-shifting water spirit from Scottish folklore, often associated with lakes, rivers, and other bodies of water throughout Scotland. Typically, Kelpies are depicted as large, powerful black horses with dripping, kelp-like manes, though they can also assume human form to deceive and lure their victims. In their horse form, Kelpies appear majestic and alluring but have an eerie, almost supernatural glow to their eyes. Legends describe their smooth, glossy coat, which conceals an underlying strength and cunning. In human form, Kelpies often appear as handsome men or women, drawing unsuspecting travelers close to the water's edge, where they reveal their true nature. Known for their strength and deceptive beauty, Kelpies are feared as dangerous creatures capable of dragging people underwater.

Origin

The origin of the Kelpie legend is deeply rooted in Scottish folklore and the cultural connection to Scotland's abundant waterways. The creature's myth likely emerged as a cautionary tale, warning people—especially children—about the perils of wandering near water alone. Kelpies are often associated with rivers, particularly the River Ness, where they are believed to dwell and guard their territory fiercely. Over time, Kelpies became linked with water horses from other European folklore, such as the Nøkk of Scandinavian mythology, creating a shared mythological theme of water-bound tricksters and dangerous spirits. These tales have persisted over generations, with each retelling amplifying the Kelpie's fearsome reputation.

Lore

In Scottish folklore, Kelpies are feared and respected as guardians of water, known for their cunning and often malevolent intent. They are said to haunt lonely stretches of water, waiting for someone to approach before transforming into a beautiful, alluring horse or human. Once lured in, victims would find themselves unable to escape, as the Kelpie's skin was believed to become adhesive, trapping anyone who touched it. With great strength, the creature would then drag its captive beneath the water to drown. While typically malevolent, some tales suggest that Kelpies can be outwitted or even tamed with the right tools, such as a bridle inscribed with a cross. However, such stories are rare.

Additional Facts
- Some believe that the legend of the Kelpie may have been inspired by real dangers associated with water, such as whirlpools or sudden currents, which could pull unsuspecting people under.
- While generally portrayed as dark and dangerous, Kelpies are occasionally depicted as protectors of their waters, punishing only those who disrespect nature.

Additional Resources

1. Scottish Myths and Legends by Donald A. Mackenzie offers a detailed exploration of water spirits in Scottish folklore.

47

KONGAMATO

(Zambia)

Description

The Kongamato is a cryptid from Zambia and surrounding regions, often described as a large, bird-like creature with bat-like wings and a fearsome, reptilian appearance. Typically, it is said to have a wingspan of four to seven feet, with leathery, membranous wings similar to those of a bat or prehistoric pterosaur. Its body is described as covered in dark, tough skin, with a long, pointed beak filled with sharp teeth. The creature's coloring is said to range from reddish-brown to black, adding to its menacing appearance. Witnesses often report seeing the Kongamato flying over rivers or swamps, particularly in remote or heavily forested areas, where it allegedly hunts for fish and other small animals. Known for its aggressive behavior, the Kongamato is reputed to attack anyone who ventures too close to its territory, adding to its fearsome reputation.

Origin

The legend of the Kongamato originates among the indigenous tribes of Zambia, Angola, and the Congo Basin. The name "Kongamato" loosely translates to "overturner of boats," a reflection of its alleged attacks on fishermen who venture too far into its territory. The creature gained attention in the Western world in the early 20th century, particularly after reports by British explorers and colonists who heard tales of the beast from locals. In some cases, Westerners even claimed to have seen the creature themselves, sparking debates and interest in the possibility of a living pterosaur-like animal. Over time, the legend has continued to intrigue cryptozoologists and researchers, drawing adventurers and explorers to Zambia in search of evidence for the elusive Kongamato.

Lore

In local Zambian folklore, the Kongamato is often regarded as a dangerous spirit of the rivers and swamps, a creature that protects its domain with ferocity. Some accounts suggest it possesses supernatural abilities, appearing only to those who disrespect or pollute the waters it inhabits. The Kongamato is said to attack intruders with its sharp claws and beak, leaving wounds that are feared to be cursed or difficult to heal. Despite its fearsome reputation, sightings of the creature are considered rare, with some locals believing that the Kongamato only appears under specific conditions or at certain times of the year.

Additional Facts
- Western explorers and naturalists of the early 20th century, such as Frank Melland, documented tales of the Kongamato in their writings, noting the strong beliefs of the local people regarding its existence.

- Cryptozoologists have compared the Kongamato to other reported pterosaur-like cryptids, such as the Ropen of Papua New Guinea, due to similarities in physical description and behavior.

Additional Resources

1. In Witchbound Africa by Frank Melland includes firsthand accounts of local beliefs about the Kongamato, capturing the fascination and fear it inspires.

48

KRAMPUS

(Austria)

Description

Krampus is a cryptid-like creature reported in the Alpine regions of Austria, often described as a terrifying, demonic figure associated with the winter season. Sightings describe him as a tall, humanoid entity with shaggy, dark fur covering his body, large, curling horns, and a long, snake-like tongue. His legs end in cloven hooves, and he is sometimes said to carry chains or a bundle of birch branches, adding to his fearsome reputation. Witnesses report a menacing, muscular build and piercing, glowing eyes that give him an unsettling, supernatural aura. Krampus is reputed to move silently despite his imposing size, adding to the mystery surrounding his appearances in isolated villages and towns, particularly during the cold winter months.

Origin

The origins of the Krampus sightings can be traced to rural Alpine folklore, where locals have long reported encountering this cryptid during the early winter season. Stories of Krampus have been passed down for centuries, with local residents attributing his appearances to an ancient guardian of the forests or a spirit of vengeance. In particular, the legend grew around the 17th century, with the creature supposedly appearing on the eve of St. Nicholas Day, when many families prepared for both blessings and possible warnings. Though often dismissed as folklore, some researchers speculate that Krampus may have been inspired by ancient sightings of an unknown cryptid or a misunderstood animal species adapted to the region's dense forests and cold climate.

Lore

In cryptid lore, Krampus is regarded as a reclusive but highly territorial creature, often said to punish those who venture too close to his domain or act disrespectfully toward nature. Sightings are most common in rural areas of Austria, where Krampus is believed to emerge from mountain caves or hidden valleys to roam near villages. Descriptions suggest that he targets homes with small children, with reports claiming that he taps on windows or leaves claw marks as a sign of his presence. Locals often describe feeling an intense sense of dread or the smell of sulfur in the air before a sighting, leading many to believe that Krampus has a supernatural or otherworldly origin. While his behavior is primarily seen as territorial, Krampus's appearances have long instilled fear and respect for the wilderness among local communities.

Additional Facts
- The pungent, sulfuric odor often reported in Krampus sightings has led some to theorize that he may inhabit volcanic areas or areas rich in geothermal activity.

- In certain Alpine regions, people leave small offerings or symbols outside their homes during the winter to appease Krampus and avoid his attention.

Additional Resources

1. The documentary Legendary Creatures of the Alps examines cryptids like Krampus.

49

LA LUZ MALA

(Argentina)

Description

La Luz Mala, meaning "The Evil Light" in Spanish, is a mysterious and eerie phenomenon reported in rural Argentina and parts of Uruguay. Described as ghostly orbs of light floating above the ground, La Luz Mala appears in remote areas, especially in the countryside, near swamps, and fields. These lights are typically white or bluish but are sometimes seen in green or red. Witnesses report that they appear suddenly and often follow those who try to approach or flee from them. La Luz Mala is known to float silently, adding an unsettling quality to its presence. Some accounts describe the lights as large, almost fireball-like, while others are small, barely illuminating the ground around them.

Origin

The legend of La Luz Mala has deep roots in Argentine folklore, with Indigenous and early colonial stories warning of ghostly lights appearing at night. In local belief, these lights are often thought to be the spirits of the dead, particularly those who did not receive a proper burial. According to lore, these lights are souls that wander restlessly, marking areas where treasure or human remains are buried. Some people interpret La Luz Mala as a spiritual warning or an omen of death. Scientists, however, suggest that the lights may be a natural phenomenon caused by the combustion of gases from decaying organic matter, yet the legend of La Luz Mala continues to thrive in Argentine culture.

Lore

La Luz Mala is often regarded as both a supernatural and cautionary tale, instilling respect for the dead and the sanctity of burial sites. Locals believe that approaching the lights is dangerous and that they should be avoided to prevent harm or misfortune. Some traditions hold that kneeling, praying, or lying face-down can help avoid the light's attention, allowing a safe escape. The lights are sometimes said to appear during significant life events or around times of illness, adding to their reputation as an omen. While stories of La Luz Mala are told across Argentina, they are particularly common in rural areas, where sightings have been passed down through generations.

Additional Facts
- Local traditions interpret La Luz Mala as a warning sign or an indication of a spirit searching for peace.
- The lights are sometimes linked to treasure hunting, as folklore suggests they appear above hidden valuables or graves.
- La Luz Mala is sometimes compared to other ghost light phenomena worldwide, such as the will-o'-the-wisp in European folklore.

Additional Resources

1. The book "Legends and Myths of Argentina" by Juan Carlos Pérez explores La Luz Mala and other supernatural folklore from Argentina.

50

LOCH NESS MONSTER

(Scotland)

Description

The Loch Ness Monster, often affectionately referred to as "Nessie," is a cryptid reputed to inhabit Loch Ness, a large, deep freshwater lake near Inverness in the Scottish Highlands. Nessie is most commonly described as a large, serpentine or plesiosaur-like creature with a long, slender neck, humps on its back, and an elusive, almost prehistoric appearance. Sightings often report a dark, smooth body that appears to ripple as it moves through the water, sometimes with a small head emerging above the surface. The creature's size is estimated to be around 20 to 40 feet in length, though reports vary widely. Known for its shy, elusive nature, the Loch Ness Monster has intrigued visitors for generations, adding an aura of mystery to the serene Scottish loch.

Origin

The Loch Ness Monster legend dates back to ancient times, with some of the earliest records appearing in 6th-century AD when Saint Columba reportedly encountered a water creature in the River Ness. The modern Nessie phenomenon began in 1933, when a couple claimed to have seen a large creature crossing the road near the loch. This encounter sparked a wave of sightings, with locals and tourists reporting glimpses of the creature in Loch Ness. The following year, the famous "Surgeon's Photograph" was published, supposedly showing Nessie's long neck and head above the water. Although this photo was later debunked as a hoax, it cemented Nessie's status as one of the most famous lake monsters in the world.

Lore

In local folklore, Nessie is often depicted as a shy, gentle giant, occasionally surfacing to reveal glimpses of its mysterious form before vanishing back into the deep. Some theories suggest that Nessie could be a surviving plesiosaur, while others propose that it might be a giant eel, a sturgeon, or even an undiscovered species. Skeptics argue that sightings of the monster could be misidentified logs, waves, or other natural phenomena. Despite the lack of conclusive evidence, Nessie has become a beloved figure in Scottish culture, symbolizing both the mystique of the Highlands and the allure of the unknown. The legend has attracted countless cryptozoologists, tourists, and researchers to Loch Ness, contributing to its global fame as a possible home for the unexplained.

Additional Facts
- The depth of Loch Ness, reaching up to 755 feet, along with its murky water, has fueled theories that such a creature could remain hidden in its depths.

- Several sonar scans and scientific investigations have been conducted in Loch Ness, yet no definitive evidence has been found to prove Nessie's existence.

Additional Resources

1. The Loch Ness Mystery Solved by Ronald Binns provides an in-depth analysis of the cryptid.

51

LOPEZ ISLAND BIGFOOT

(Washington, USA)

Description

The Lopez Island Bigfoot is a cryptid reportedly seen on Lopez Island in Washington's San Juan Islands, believed to be a variant of the traditional Bigfoot or Sasquatch. Descriptions characterize it as a massive, ape-like being standing between seven and nine feet tall, covered in thick, dark fur that blends into the dense, forested surroundings. Witnesses describe its muscular build, broad shoulders, and distinctively large footprints left behind in muddy, secluded areas. The creature's face is said to be a mix between human and ape, with deep-set eyes that convey an intense awareness of its surroundings. Known for its stealth and elusiveness, the Lopez Island Bigfoot is rarely sighted, yet those who have encountered it report an overwhelming sense of awe and dread.

Origin

Sightings of the Lopez Island Bigfoot are relatively recent compared to traditional Sasquatch legends, first emerging in local reports during the 20th century. However, the indigenous Coast Salish people have long held beliefs in "wild men" or "stick Indians," mysterious beings inhabiting the remote areas of the Pacific Northwest. While the Lopez Island Bigfoot legend is distinct from these traditional tales, it shares similarities in descriptions and characteristics. The dense, mist-covered forests of Lopez Island, with limited human presence and many inaccessible areas, are seen as an ideal habitat for such a creature, fueling speculation that a reclusive, undiscovered being could inhabit the area.

Lore

Local lore portrays the Lopez Island Bigfoot as a guardian of the forest, deeply connected to the wilderness. Many believe the creature only reveals itself to a select few who unintentionally wander into its territory, often leaving them with a profound sense of awe or fear. Stories suggest that it can move silently through the woods and disappear without a trace, leaving behind only footprints or broken branches. Encounters are usually brief, with the Bigfoot quickly retreating into the forest upon noticing human presence. Though sightings are rare, the Lopez Island Bigfoot has become a mysterious figure in local folklore, often discussed among residents and cryptid enthusiasts who visit the island hoping for evidence of its existence.

Additional Facts
- The Lopez Island Bigfoot has drawn comparisons to other isolated Bigfoot populations, such as the Lake Worth Monster in Texas and the Florida Skunk Ape, though it is known for its more peaceful, reclusive nature.

- Certain residents have organized unofficial "Bigfoot watch" groups, exploring remote areas of the island in hopes of capturing evidence of the creature.

Additional Resources

1. Bigfoot in the Pacific Northwest by David George Gordon explores Bigfoot sightings across the region, including tales from Washington's San Juan Islands.

52

LOVELAND FROGMAN

(Ohio, USA)

Description

The Loveland Frogman, also known as the Loveland Lizard, is a cryptid reported to inhabit the small town of Loveland, Ohio. Described as a large, humanoid frog or lizard, this creature reportedly stands about 3 to 4 feet tall with leathery, gray-green skin and a distinctly frog-like face. Witnesses describe it as having wide, bulbous eyes, long limbs, and webbed hands and feet, which enable it to move both on land and in water. The Loveland Frogman is said to emit a strong odor and, in some accounts, is capable of emitting a low growl or croak. Its nocturnal nature and elusive behavior have contributed to its eerie reputation in Ohio folklore.

Origin

The legend of the Loveland Frogman dates back to 1955, when a traveling businessman reported seeing three humanoid, frog-like creatures under a bridge near the Little Miami River in Loveland. The creatures were said to be communicating with each other before one lifted a "wand" emitting sparks. The story resurfaced in 1972 when a Loveland police officer reported encountering a similar creature on the side of the road. These sightings generated local curiosity and transformed the Frogman into a regional legend. Though the sightings were met with skepticism, the creature became a subject of fascination and fear.

Lore

In Ohio folklore, the Loveland Frogman is seen as a mysterious and elusive creature, often associated with the wetlands and rivers of the Loveland area. Locals claim the creature is reclusive, only revealing itself at night and swiftly vanishing into the water if disturbed. Some legends suggest that the Frogman is part of an ancient species or even a supernatural being guarding the waterways. While some view the creature as a harmless oddity, others believe it has a more sinister side, warning travelers to avoid the riverside at night. The Loveland Frogman has since become a beloved, albeit unsettling, figure in local culture, inspiring stories, art, and even occasional festivals.

Additional Facts
- The 1972 police encounter with the Frogman remains one of the most well-documented sightings, though it was later dismissed by some as a possible hoax.

- The Loveland Frogman has appeared in various Ohio folklore collections, cementing its status as one of the region's most distinctive cryptids.

- The creature has inspired local events, including a Loveland Frogman 5K, where participants run through the Frogman's reported territory.

Additional Resources

1. Mysterious Creatures: A Guide to Cryptozoology by George M. Eberhart features an entry on the Loveland Frogman and other cryptids from around the world.

53

LUSCA

(Bahamas)

Description

The Lusca is a cryptid said to inhabit the underwater cave systems and blue holes of the Bahamas, particularly around Andros Island. Often described as a monstrous creature with elements of both an octopus and a shark, the Lusca is reputed to reach terrifying sizes, with tentacles spanning up to 75 feet or more. Witnesses report that it has a massive, elongated body with powerful, razor-sharp suckers on its tentacles and a head resembling that of a giant squid or a fearsome sea monster. The creature's skin is typically described as mottled or dark, blending seamlessly with the ocean depths and rocky cave walls, allowing it to ambush unsuspecting prey. The Lusca is infamous for its predatory nature, allegedly capable of dragging large fish, animals, and even boats beneath the surface.

Origin

The legend of the Lusca is rooted in Bahamian folklore and maritime tales that span centuries. The deep blue holes and underwater caverns of the Bahamas, which are geographically unique and notoriously dangerous, have long fueled stories of lurking monsters. Local fishermen and divers have shared stories of strange sightings and unexplained disappearances, contributing to the Lusca's fearsome reputation. Some believe that the Lusca could be a surviving giant cephalopod or an undiscovered species adapted to the unique conditions of the blue holes. The creature's legend has persisted, gaining attention from both locals and cryptid enthusiasts worldwide.

Lore

In Bahamian lore, the Lusca is often portrayed as a guardian of the blue holes, defending its hidden domain against intruders. The creature is said to be highly territorial and aggressive, capable of causing whirlpools or sudden currents to pull boats into its lair. Some stories warn that the Lusca can move swiftly between underwater tunnels, making it impossible to track or escape once it has set its sights on its target. Local legends also suggest that it may be drawn to movement or noise in the water, making it especially dangerous to divers and fishermen who unknowingly wander too close. The Lusca is both feared and respected as a powerful force of nature, symbolizing the mysterious and untamed spirit of the sea.

Additional Facts
- Local myths suggest that the Lusca can create whirlpools or underwater disturbances, contributing to its legend as a fierce and otherworldly predator.
- The Lusca's reputation has spread beyond the Bahamas, with stories and documentaries exploring its myth and possible existence, sparking interest in cryptozoology communities.

Additional Resources

1. Blue Holes of the Bahamas by Brett Hemphill provides an exploration of the natural wonders of the Bahamian blue holes, including references to the Lusca legend.

54

MANTICORE

(Persia)

Description

The Manticore is a legendary creature from ancient Persian mythology, known for its fearsome appearance and predatory nature. Described as a blend of lion, human, and scorpion, the Manticore features the body of a large lion with a muscular frame, a human-like face with sharp, carnivorous teeth, and a long, spiked tail. The tail is often depicted with venomous barbs or spines capable of launching at prey or intruders, making it both ferocious and highly dangerous. Some accounts suggest the Manticore has wings, giving it an even more terrifying presence as it prowls for prey. Known for its blood-red fur and hypnotic, piercing gaze, the Manticore is said to emit a haunting, almost human-like roar that echoes across the lands.

Origin

The Manticore legend originates from ancient Persia, where it was a creature of folklore and myth, symbolizing both danger and mystery. Its name is derived from the Old Persian term martiya khvara, meaning man-eater, a fitting title for a beast said to devour humans whole, leaving no trace behind. Early Persian texts mention the Manticore as a creature feared by travelers and warriors alike, warning of its deadly attacks in remote wildernesses. Greek writers, like Ctesias, introduced the Manticore to Western audiences, who were fascinated by this exotic and terrifying beast. The Manticore continued to appear in medieval bestiaries and tales across Asia and Europe, becoming a symbol of untamed power and lethal cunning.

Lore

In Persian lore, the Manticore is often depicted as an insatiable predator, guarding remote areas and hidden treasures. It is known to lie in wait, ambushing travelers, and vanishing back into the wilds without a trace. Many tales warn that the Manticore possesses an uncanny intelligence, able to mimic human speech to lure its prey closer. Stories of the Manticore speak of its role as both a guardian of forbidden lands and a terrifying force of nature, symbolizing humanity's fear of the unknown. Medieval legends expanded on the Manticore's lore, adding elements of shape-shifting and magical resilience, further enhancing its fearsome reputation.

Additional Facts
- The Manticore's ability to mimic human speech has led to associations with trickery and deceit in various folklore interpretations.

- In Persian art and literature, the Manticore was often depicted as a symbol of destruction and resilience, representing the dangers that lurk beyond civilization's reach.

Additional Resources

1. The Persian Myths by Vesta Sarkhosh Curtis explores the Manticore and other creatures from ancient Persian folklore, delving into their cultural significance.

55

MAPINGUARI

(Brazil)

Description

The Mapinguari is a cryptid from the Amazon rainforest in Brazil, often described as a towering, ape-like creature or a giant sloth-like being with fearsome features. Accounts vary, but witnesses commonly describe the Mapinguari as standing around seven feet tall, covered in reddish or dark fur, and possessing long, sharp claws. Its most distinctive feature is a second mouth on its belly, reportedly capable of emitting a foul odor that repels predators and intruders. Some descriptions suggest it has a single eye in the center of its forehead or an unusual gait, adding to its mysterious, otherworldly appearance. Known for its deep growls and pungent scent, the Mapinguari is said to roam the dense jungles and mountains, particularly in remote areas far from human settlements.

Origin

The Mapinguari legend has deep roots in Amazonian folklore, where indigenous tribes regard it as a guardian of the forest. The creature is believed to protect the jungle, punishing those who exploit its resources or disrespect its delicate ecosystem. Some local tribes see the Mapinguari as a spirit or an ancient being connected to the mysteries of the Amazon. While many believe the creature is purely mythical, some speculate that the legend could stem from memories of prehistoric species like the giant ground sloth, which once inhabited the region.

Lore

In Amazonian lore, the Mapinguari is both feared and respected as a protector of the rainforest. Stories describe it emerging from the depths of the jungle to punish poachers or others who harm the forest. Its pungent stench is said to signal its approach, giving people a chance to flee before encountering its imposing form. The Mapinguari is also rumored to possess nearly supernatural resistance to harm, with skin that repels spears and bullets. Those who claim to have survived encounters often recall feeling an intense dread, with the creature's powerful stench lingering long after it has disappeared back into the forest. This fierce guardian serves as a reminder of the mystery and power of the Amazon, urging people to tread carefully in its domain.

Additional Facts
- The creature's eerie, human-like cries and loud growls are described as echoing through the jungle, adding to the fear felt by those who encounter it.

- The Mapinguari has attracted the attention of cryptozoologists, and occasional expeditions venture into the Amazon to search for evidence of its existence.

Additional Resources

1. Amazon Beasts and Legends by Roderick Lewis explores Amazonian cryptids, including the Mapinguari and other mythical creatures of the rainforest.

56

MAROZI

(Kenya)

Description

The Marozi, often called the "Spotted Lion," is a cryptid reportedly sighted in the highlands of Kenya, particularly in the Aberdare Mountains. Unlike typical African lions, the Marozi is distinguished by its unique appearance, most notably its leopard-like spots. Witnesses describe it as a medium-sized feline, resembling a lion but with smaller, more compact features and distinct dark spots or rosettes along its coat. Unlike juvenile lions, which may show faint spots that fade with maturity, the Marozi's markings are said to persist into adulthood, giving it a hybrid look between a lion and a leopard. Reports also suggest that Marozi tend to travel in small groups and are rarely seen near human settlements, which adds to their mysterious and elusive reputation.

Origin

The first documented encounter with the Marozi came in the early 20th century when British hunters in Kenya reported sightings of spotted lions. In 1931, a pair of spotted lion pelts was discovered in the Aberdare Mountains, leading to increased interest and speculation about whether a subspecies of lion, distinct from the common African lion, might inhabit the region. Various theories emerged, with some suggesting that the Marozi could be a natural hybrid between lions and leopards, while others believe it might be an undiscovered subspecies adapted to the highland environment. Despite efforts by researchers and explorers, no conclusive evidence has confirmed the Marozi's existence, but local legends and occasional sightings continue to keep the mystery alive.

Lore

In Kenyan folklore, the Marozi is often seen as a guardian of the highlands, watching over the rugged terrain where few other large predators dare to roam. Some stories suggest that the Marozi are solitary protectors of remote mountain passes and valleys, steering clear of humans but fiercely defending their domain from other predators. Local communities consider the Marozi both a natural wonder and a symbol of Kenya's wild landscapes, respecting its reclusive nature. Encounters with the Marozi are rare and brief, with the creature reportedly vanishing into dense cover when spotted. Its adaptability to high-altitude environments and its unique spotted appearance have given it a legendary status among locals and cryptid enthusiasts alike.

Additional Facts

- The two pelts discovered in 1931 remain one of the most substantial pieces of evidence for the Marozi's existence, although no physical specimen has been captured or observed since.

- In local Kenyan art and storytelling, the Marozi is sometimes depicted alongside other cryptids and wildlife as a unique symbol of Kenya's biodiversity.

Additional Resources

1. The book Mystery Cats of the World by Karl Shuker includes a detailed section on the Marozi.

57

MELON HEADS

(Ohio, USA)

Description

The Melon Heads are cryptid-like beings from the folklore of Ohio, particularly in the northeastern region around Kirtland and Chardon. They are described as small, humanoid creatures with unusually large, bulbous heads and thin, emaciated bodies. Witnesses report that the Melon Heads have exaggerated, often grotesque facial features, including wide eyes and long, bony fingers. Their pale skin is said to look unnatural, and some accounts mention that they wear ragged clothing, hinting at a reclusive, feral existence. Known for their eerie appearances at night, Melon Heads are typically sighted near dense forests, isolated back roads, and abandoned structures.

Origin

The legend of the Melon Heads in Ohio is believed to date back to the mid-20th century, though some elements of the story may be older. According to local lore, the Melon Heads were once children suffering from a medical condition that caused their heads to grow abnormally large. In one version of the tale, they were mistreated and experimented upon in a secret asylum or by a rogue doctor, leading to their deformities and reclusive lifestyle. After escaping, they supposedly sought refuge in the forests and became hostile to anyone who ventured into their territory. The exact origins of the story vary, with some suggesting that it was created as a cautionary tale to discourage children from wandering too far into the woods.

Lore

In Ohio's folklore, the Melon Heads are both feared and pitied figures. Many local tales portray them as territorial and sometimes aggressive, warning travelers to avoid specific back roads or forest trails where they are known to roam. Encounters often involve strange sounds, sightings of small figures lurking in the shadows, or even brief chases through the woods. Despite their reputation, some versions of the legend paint the Melon Heads as tragic figures, driven to isolation due to their appearance and the traumatic experiences they endured. Their presence in Ohio folklore has inspired local legends, ghost tours, and continues to draw curious explorers to the region.

Additional Facts
- Local residents and visitors often report strange sounds and eerie feelings when traveling on certain roads associated with Melon Head sightings, particularly at night.
- The Melon Heads have inspired local horror stories, urban legends, and have even been referenced in popular culture, particularly in regional ghost stories and folklore literature.

Additional Resources

1. The book Weird Ohio: Your Travel Guide to Ohio's Local Legends and Best Kept Secrets covers stories of the Melon Heads and other unusual Ohio folklore.

2. American Myths, Legends, and Tall Tales: An Encyclopedia of American Folklore.

58

MENEHUNE

(Hawaii, USA)

Description

The Menehune are small, humanoid beings from Hawaiian folklore, often depicted as skilled and industrious people who inhabit the dense forests and hidden valleys of the Hawaiian Islands. Standing between two to three feet tall, they are known for their stocky builds, dark skin, and agile, nimble bodies adapted for navigating rugged terrain. Legends describe them as master builders, credited with constructing impressive structures like fishponds, temples, and roads—often completing these projects overnight. Despite their small stature, Menehune possess incredible strength and are known for their intelligence and craftsmanship. They are also associated with a mischievous, reclusive nature, only appearing to those who respect the land or those of pure Hawaiian lineage.

Origin

The legend of the Menehune is deeply rooted in Hawaiian mythology, believed by some to predate the arrival of Polynesian settlers. According to local traditions, the Menehune were an ancient people who once lived throughout the islands but eventually retreated into the mountains and forests as other tribes populated the lowlands. Stories of their existence are recorded in oral histories passed down through generations. Some anthropologists theorize that these tales may have originated as references to an early race of Hawaiian settlers, while others believe that the Menehune serve as symbols of a connection between the Hawaiian people and the natural world.

Lore

In Hawaiian lore, the Menehune are both respected and feared for their skills and elusive behavior. They are often said to possess magical abilities, allowing them to work quickly and avoid detection. Many legends speak of the Menehune constructing fishponds, aqueducts, and even entire villages overnight, disappearing before dawn to avoid being seen. If a Menehune is interrupted or witnessed during their work, it is believed they will abandon the project entirely, leaving it unfinished. They are also known for playing tricks on those who disrespect the land or disrupt their hidden way of life. However, those who show reverence to the Menehune and the natural world may receive their help, particularly in agricultural and fishing endeavors.

Additional Facts
- The Menehune are often compared to European fairies or dwarves due to their small stature and talent for craft, though they remain distinct to Hawaiian culture and mythology.
- Sightings and tales of Menehune occasionally surface, with locals sometimes attributing unexplained nighttime sounds or small stone structures to these elusive beings.

Additional Resources
1. The book Hawaiian Legends of the Menehune by Mary Kawena Pukui provides a detailed look at traditional stories and beliefs surrounding the Menehune.

59

MERMAID

(GLOBAL FOLKLORE)

Description

Mermaids are cryptid sea creatures reported across coastal regions worldwide, often depicted as part human and part fish, with a humanoid torso, arms, and head merging seamlessly into a long, fish-like tail. Unlike the glamorous, mystical mermaids of folklore, cryptid sightings portray them as elusive, sometimes eerie beings with slippery skin, webbed hands, and sharp, piercing eyes adapted to the depths of the ocean. Witnesses report seeing mermaids with varying degrees of beauty and strangeness, with some having scales and darker, seaweed-like hair that blends into the waves. These creatures are known for their stealth and tend to avoid human contact, often seen in shadowy waters or as fleeting glimpses at a distance before vanishing into the depths.

Origin

Mermaid sightings have been reported for centuries, with accounts dating back to early European explorers who believed they encountered mermaids on the open sea. Christopher Columbus even recorded seeing "female forms" in the waters near the Caribbean, describing them as less beautiful than expected and more humanoid and fish-like. In the 19th and 20th centuries, sailors and fishermen reported seeing mermaid-like figures off the coasts of various countries, particularly in areas like the Caribbean, the Mediterranean, and the cold waters of Northern Europe. While some stories may have been inspired by manatees or seals, the consistent details of mermaid sightings worldwide add a cryptid element to their legend, hinting at something unknown lurking beneath the waves.

Lore

Cryptid lore surrounding mermaids often presents them as ambivalent or even dangerous beings. Many accounts warn of their tendency to lure sailors closer to perilous reefs or rocky shorelines. Some tales describe mermaids as guardians of the sea who become hostile when humans encroach on their waters. Fishermen sometimes claim to hear their eerie singing or splashing just before they vanish beneath the surface, leaving nothing but ripples in the water. Unlike mythological depictions, these cryptid mermaids are more solitary and reclusive, often appearing only in remote or rugged coastal areas. While cryptid mermaids are often described as cautious and elusive, some stories depict them as highly intelligent creatures capable of understanding human presence and avoiding detection.

Additional Facts
- In modern times, mermaid sightings are rare, but reports continue to surface occasionally, particularly in coastal towns with strong fishing traditions.
- Mermaid cryptid sightings share similarities with other aquatic cryptids, such as the Ningen in Japan or the Sirena of the Philippines, suggesting a possible global phenomenon.

Additional Resources

1. Cryptids of the Sea by Richard Freeman delves into aquatic cryptids, including mermaids and other mysterious creatures reported by sailors.

60

MINIWAKEE

(Wisconsin, USA)

Description

The Miniwakee is a mysterious cryptid reportedly lurking in the dense forests and remote areas of northern Wisconsin, USA. Often described as a small, elusive creature with an unusual appearance, the Miniwakee is said to stand between two and three feet tall, with a humanoid shape. Witnesses describe it as having a wrinkled, leathery face, elongated arms, and large, intelligent eyes that glow faintly in low light. Its skin is often depicted as dark and rough, camouflaging it in shadowed, wooded areas. The Miniwakee moves quietly and quickly, making it difficult to track, and is known for its eerie vocalizations, which range from quiet murmurs to unsettling shrieks. This cryptid's ability to vanish into the underbrush adds to its air of mystery, leaving behind only faint tracks or broken branches as evidence of its presence.

Origin

The legend of the Miniwakee is rooted in local Native American folklore, where similar creatures have been described as forest spirits or protectors of the land. Early settlers in the region reported strange sightings of small, humanoid beings flitting through the trees or observing them from afar. Over time, these accounts evolved into stories of the Miniwakee as a cryptid, often associated with the untamed wilderness of Wisconsin's forests. Locals believe that the Miniwakee is a solitary creature, guarding its territory from intruders and remaining hidden from human view.

Lore

In local lore, the Miniwakee is seen as a symbol of the wild, untamed spirit of Wisconsin's northern forests. Some consider it a benign creature, acting as a guardian of the forest, while others see it as a trickster, luring unsuspecting travelers off paths and into the woods. Sightings are rare, and many believe that encountering a Miniwakee is a sign to respect the forest and tread carefully. Some stories suggest that it has supernatural abilities, such as blending seamlessly into its surroundings or communicating through strange sounds. The Miniwakee's legend has sparked curiosity and intrigue, leading hikers and cryptid enthusiasts to search for signs of this elusive creature in the shadows of Wisconsin's woods.

Additional Facts
- The Miniwakee has inspired local legends and tourist interest, with people traveling to northern Wisconsin in hopes of glimpsing this elusive creature.

- Despite its frightening appearance, the Miniwakee is often described as reclusive and non-threatening, choosing flight over fight in encounters with humans.

Additional Resources

1. The Cryptids of Wisconsin by David Weatherly explores Wisconsin's cryptid lore, with a section on the Miniwakee and its impact on local culture.

61

MOKELE-MBEMBE

(Congo)

Description

The Mokele-Mbembe is a cryptid said to inhabit the remote swamps, rivers, and lakes of the Congo Basin in Central Africa. Often described as a massive, dinosaur-like creature, the Mokele-Mbembe has drawn comparisons to sauropod dinosaurs like the Apatosaurus, given its reported long neck, bulky body, and long tail. Witnesses claim it has smooth, dark brown or grayish skin, with a size reaching up to 30 feet or more in length. Mokele-Mbembe is generally depicted as a herbivorous creature, feeding on vegetation along the water's edge, though some accounts suggest it can be territorial and may defend its domain aggressively. The creature's elusive behavior, combined with the dense, uncharted wilderness of the Congo, has kept it shrouded in mystery.

Origin

The legend of Mokele-Mbembe originates from the oral traditions of the indigenous people of the Congo Basin, who describe it as a powerful and revered animal that has roamed the rivers and swamps for centuries. The name "Mokele-Mbembe" itself roughly translates to "one who stops the flow of rivers," highlighting its connection to water and its purported size. Western explorers and missionaries first learned of the creature in the early 20th century, when local guides and villagers shared stories about it, sparking interest in the possibility of a living dinosaur. Over the years, expeditions have attempted to find evidence of the Mokele-Mbembe, with some claiming to have seen tracks or heard its calls, though no definitive evidence has ever been found.

Lore

In local Congolese lore, Mokele-Mbembe is often considered a spirit of the river, a creature with a profound presence that must be respected. Some tribes view it as a protector of the land, while others believe encountering it is an omen or a warning. Stories say that Mokele-Mbembe has been known to defend its territory against fishermen or hunters who encroach on its waters, and some tales recount instances where villagers lost their lives after crossing paths with the creature. Despite the potential dangers, Mokele-Mbembe holds a deep cultural significance, representing both the power of the natural world and the mysteries that remain hidden within the Congo's vast wilderness.

Additional Facts
- Local legends and stories are central to the creature's identity, with the Mokele-Mbembe inspiring cautionary tales about respecting nature and protecting the ecosystem.

- The creature has been the subject of numerous documentaries, books, and cryptid research, capturing the imaginations of those fascinated by the possibility of "living dinosaurs."

Additional Resources

1. Cryptozoology: A to Z by Loren Coleman and Jerome Clark offers an overview of cryptids worldwide, including the Mokele-Mbembe.

62

MONGOLIAN DEATH WORM

(Mongolia)

Description

The Mongolian Death Worm, known locally as olgoi-khorkhoi or "large intestine worm," is a cryptid reputed to inhabit the remote sands of the Gobi Desert in Mongolia. Descriptions of the creature suggest it is a large, worm-like being, typically between two to five feet long, with a cylindrical, reddish body reminiscent of an earthworm or snake. Witnesses claim it has a smooth, segmented surface and no discernible eyes or mouth, yet it moves with surprising speed beneath or across the sand. Its striking red color and alleged ability to deliver deadly electric shocks or emit a corrosive venom that paralyzes or kills its prey add to its fearsome reputation.

Origin

The legend of the Mongolian Death Worm is deeply rooted in Mongolian folklore and was first brought to international attention in the early 20th century by explorer Roy Chapman Andrews. Mongolian nomads and desert dwellers tell stories of this cryptid, warning travelers and locals of the dangers of encountering it. These stories often involve fatal encounters, where the worm supposedly emerges from the sand, delivering a sudden, lethal attack before retreating into the dunes. Its association with death and venom has made it a figure of caution in local culture and a source of intrigue for cryptid researchers.

Lore

According to local legends, the Mongolian Death Worm is a reclusive creature that only surfaces during the hottest months of the year. It is said to appear suddenly, attacking animals, livestock, and even humans with either a potent electrical shock or its supposed venomous spray. Some say the worm has a supernatural quality, with stories describing it as a guardian of ancient treasures buried beneath the desert. Other accounts suggest that the creature only attacks when threatened, staying hidden deep in the sands for most of the year. The fear surrounding the Mongolian Death Worm has cemented it as one of the most infamous and dangerous cryptids in Central Asian folklore.

Additional Facts
- Reports of the creature emitting deadly venom or electric shocks have led some scientists to believe it might be based on exaggerated encounters with venomous or toxic desert animals.
- Despite numerous expeditions in search of the Mongolian Death Worm, no physical evidence, such as bones or remains, has ever been found, adding to its legendary status.
- The Death Worm legend has become a popular subject in media, inspiring books, movies, and TV shows that explore the mystery and danger surrounding the creature.

Additional Resources

1. The book On the Trail of Ancient Man by Roy Chapman Andrews explores early accounts of the Mongolian Death Worm from the Gobi Desert.

63

MOTHMAN

(West Virginia, USA)

Description

The Mothman is a legendary cryptid reported in the town of Point Pleasant, West Virginia. Described as a large, humanoid figure with enormous wings and glowing red eyes, the Mothman has become a subject of fascination since its first reported sighting in the mid-1960s. Witnesses commonly describe it as standing between six and seven feet tall with a wingspan of up to ten feet, its eyes emitting an intense, almost hypnotic red glow. The creature's wings are typically said to resemble those of a bat or moth, with its body covered in dark, feather-like or leathery textures. The Mothman is known for its silent, ominous flight and the eerie sense of dread that many witnesses report feeling in its presence.

Origin

The Mothman legend began in November 1966, when two young couples reported seeing a "man-sized bird" with glowing eyes near an abandoned munitions factory in Point Pleasant. Over the following year, additional sightings occurred, and the creature became the center of intense media attention, captivating the nation. The string of Mothman encounters coincided with other strange events in the area, including reports of UFOs and mysterious lights in the sky. The Mothman legend reached its peak with the tragic collapse of the Silver Bridge in December 1967, leading some to believe that the creature's appearance was an omen of disaster. This connection cemented the Mothman's place in local folklore and added a supernatural dimension to the story, though no conclusive evidence of its existence was ever found.

Lore

In West Virginian lore, the Mothman is often viewed as a harbinger of doom or a supernatural warning sign, its appearance linked to impending misfortune or tragedy. Some locals believe it to be an interdimensional being, an alien, or a guardian spirit attempting to warn humanity. Others see it as a malevolent creature drawn to chaotic or sorrowful events. Stories surrounding the Mothman continue to evolve, with occasional sightings and reported experiences fueling theories that it still haunts the region. The creature has since become a cultural symbol, inspiring books, movies, and annual festivals, with many viewing it as both a chilling and intriguing part of West Virginian identity.

Additional Facts
- The creature is often associated with the Silver Bridge collapse, where 46 people tragically lost their lives, deepening its ominous reputation.
- In 2002, the Mothman Prophecies movie, based on John Keel's book of the same name, brought renewed interest in the creature, cementing its place in pop culture.

Additional Resources

1. The book The Mothman Prophecies by John A. Keel explores the strange events surrounding the Mothman.

64

MUGWUMP

Canada)

Description

The Mugwump is a cryptid from the folklore of Canada, specifically reported around the remote lakes and forested regions of Ontario and Quebec. This creature is often described as a large, aquatic being with a serpentine body, measuring up to 15 feet in length. It is said to have an elongated, flexible neck, a smooth, dark-skinned body, and a head resembling a cross between a fish and a lizard, complete with bulging eyes and a wide mouth lined with sharp teeth. Mugwumps are known to be elusive, rarely surfacing, and are often spotted by boaters and fishermen in the early hours or at dusk, moving swiftly through the water. Witnesses describe its movements as sinuous and silent, with the creature's back occasionally visible as humps rolling along the water's surface.

Origin

The Mugwump legend originates in the Indigenous and settler communities near the Great Lakes and has long been a staple of local lore. Early settlers and Indigenous communities shared stories of mysterious, serpentine beings living in the lakes, with the creature's name "Mugwump" reportedly derived from Algonquin terms related to strange aquatic animals. This cryptid legend was popularized in the mid-20th century, as tourists and locals alike reported strange sightings in Canadian lakes. Although the origins of the Mugwump are steeped in folklore, its continued sightings have led some cryptozoologists to investigate the possibility of a rare, undiscovered species dwelling in these remote waters.

Lore

In Canadian lore, the Mugwump is often viewed as a reclusive, possibly territorial creature, guarding its hidden habitat in deep, isolated lakes. Local stories describe it as a shy yet occasionally curious being, sometimes approaching boats out of curiosity before vanishing underwater. Some tales caution against angering the Mugwump, as it is rumored to capsize small boats or frighten away those who come too close. The creature's rare appearances are sometimes attributed to changes in water temperature or seasonal patterns, with most sightings occurring during warmer months. Despite its intimidating appearance, the Mugwump is not known to pose any real threat to humans, though it has become a fascinating and mysterious figure in Canadian folklore.

Additional Facts

- Some believe the Mugwump could be a relic from prehistoric times, possibly related to an ancient aquatic species, though no concrete evidence has been found.

- The creature is sometimes compared to other lake cryptids like Ogopogo or the Loch Ness Monster, yet it is distinctive for its lizard-like head and serpentine body.

Additional Resources

1. Canadian Monsters and Cryptids by Gerry Bailey delves into Mugwump lore along with other aquatic cryptids.

65

MUNGO MAN

(Australia)

Description

Mungo Man is an ancient human ancestor discovered in the Lake Mungo region of New South Wales, Australia. Believed to be around 42,000 years old, Mungo Man represents one of the oldest and most complete Homo sapiens remains found outside Africa. He was approximately 50 years old at the time of his death, with a slender build adapted to the unique environment of ancient Australia. The skeletal remains were buried with red ochre, indicating a deliberate and possibly ceremonial burial.

Origin

The remains of Mungo Man were discovered in 1974 by anthropologist Dr. Jim Bowler in the Willandra Lakes region, an area once filled with rivers and lakes. Mungo Man's discovery dramatically altered our understanding of early human migration, suggesting that humans arrived in Australia tens of thousands of years ago, much earlier than previously thought. His existence supports theories of ancient migration from Africa to Southeast Asia, and then to Australia. This ancient burial ground is part of the Willandra Lakes, a UNESCO World Heritage Site recognized for its contribution to understanding early human history.

Lore

For Indigenous Australians, Mungo Man holds profound cultural significance as a direct link to their ancient ancestors and a testament to their people's deep connection to the land. The red ochre burial—one of the earliest examples of ritual burial in human history—suggests that ancient Australians had developed spiritual and cultural practices thousands of years ago. The discovery of Mungo Man emphasized the rich history of Australia's Indigenous peoples, highlighting ancient practices of reverence for the dead. In 2017, following decades of Indigenous advocacy, Mungo Man's remains were ceremonially reburied in the Willandra Lakes region, respecting his heritage and cultural significance.

Additional Facts
- Mungo Man's bones show signs of osteoarthritis, suggesting he endured physical labor during his lifetime.

- The Lake Mungo area is also home to "Mungo Lady," another set of ancient remains, adding to the historical depth of the site.

- The Willandra Lakes region, where Mungo Man was found, is a UNESCO World Heritage site due to its importance in understanding early human life in Australia.

Additional Resources

1. The Australian Museum provides exhibits and resources on Mungo Man, showcasing his significance in Australia's ancient history.

66

MUNYANGO

(Uganda)

Description

The Munyango is a cryptid from Ugandan folklore, often described as a semi-aquatic creature that inhabits the dense swamps and riverbanks of Uganda. This mysterious being is typically reported to resemble a large, otter-like animal, measuring between five to eight feet in length. Witnesses describe the Munyango as having a muscular, streamlined body covered in dark, glossy fur, with limbs ending in webbed, clawed feet. Its head is said to be broad and flat, with a long snout, sharp teeth, and intense eyes adapted for low light. Known for its stealth and agility in water, the Munyango is believed to hunt fish, small mammals, and birds, moving swiftly through rivers and swamps.

Origin

The legend of the Munyango originates from local stories told by the native people of Uganda, particularly near Lake Victoria and other freshwater bodies. Passed down through generations, tales of the Munyango often depict it as a rare, elusive creature that appears only to those who venture too close to its territory. Some folklorists believe the Munyango legend may be inspired by encounters with unknown or rarely seen animal species in Uganda's lush and expansive wetlands.

Lore

In Ugandan folklore, the Munyango is often regarded with a mix of respect and fear. Local stories portray it as a territorial creature that defends its domain fiercely, particularly against intruders. While generally reclusive, it is known to become aggressive if provoked or threatened. Some stories suggest the Munyango possesses a supernatural quality, vanishing swiftly into the water and leaving no trace. It is considered a symbol of the hidden mysteries within Uganda's wilderness, embodying the untamed and uncharted aspects of nature.

Additional Facts

- The Munyango is often compared to cryptids from other cultures, such as the Loch Ness Monster, due to its mysterious and elusive nature.

- Some biologists theorize that the Munyango legend may be based on undiscovered or extinct species of large otters or similar mammals native to the region.

- In Ugandan art and storytelling, the Munyango is occasionally depicted as a guardian of rivers, warning locals to respect water sources and natural habitats.

Additional Resources

1. African Cryptids and Legends by John Thorpe explores lesser-known cryptids across Africa, including the Munyango.

67

NAITAKA

(Canada)

Description

The Naitaka, also known as the "Ogopogo," is a cryptid believed to inhabit Okanagan Lake in British Columbia, Canada. Often described as a massive, serpentine creature, the Naitaka is said to reach lengths of 20 to 50 feet, with dark, smooth, and scaly skin. Witnesses frequently describe it as having several humps visible above the water when it surfaces, and a head that some say resembles a horse, snake, or goat. Known for its swift, undulating movements, the Naitaka is reputed to glide gracefully across the lake, occasionally diving and reemerging, adding to its mysterious and elusive reputation.

Origin

The Naitaka legend originates from the indigenous Syilx people, who have long considered the creature a sacred guardian of Okanagan Lake. For centuries, local lore warned of the creature's formidable nature, and stories suggest that the Syilx would offer sacrifices before crossing the lake to ensure safe passage. European settlers first heard of the Naitaka in the 1800s, and since then, numerous sightings have contributed to its legendary status in Canadian folklore.

Lore

Over the years, the Naitaka has been described as both a guardian and a mysterious lake-dweller. Many consider it a warning spirit, urging respect for the natural environment. Sightings often occur on foggy days or at dusk, with witnesses claiming the creature exudes a calm yet eerie presence. Some tales describe the Naitaka as protective, only appearing to those who disturb the lake or act disrespectfully toward it. Stories of eerie sounds echoing across the water and strange ripples in otherwise still conditions contribute to the creature's lore as an elusive and cryptic being.

Additional Facts

- The creature's name, "Naitaka," means "Lake Demon" in the Syilx language, underscoring its revered status in local indigenous culture.
- Okanagan Lake's unusual depth and size contribute to speculation about unknown creatures possibly residing in its depths.
- Modern tourism in the Okanagan Valley celebrates the Naitaka with attractions, statues, and events, making it a cultural icon of the region.

Additional Resources

1. In Search of Lake Monsters by Peter Costello examines various lake cryptids, including the Naitaka.
2. The Okanagan Heritage Museum in British Columbia includes exhibits on the Naitaka, showcasing historical sightings and cultural significance.

68

NAKULA

(Alaska, USA)

Description

The Nakula is a mysterious cryptid from the folklore of Alaska, often described as a large, bear-like creature adapted to the frigid, remote wilderness. Sightings of the Nakula characterize it as a hulking beast with thick, dark fur, standing anywhere from eight to ten feet tall on its hind legs. Witnesses report a muscular build and powerful limbs, with clawed feet adapted for gripping icy terrain. Its face is said to resemble a blend between a bear and a wolf, featuring sharp, predator-like eyes and a long snout filled with powerful teeth. The Nakula is noted for its elusive nature, with sightings mainly occurring in isolated, forested, or mountainous areas during the night or twilight hours. Some accounts describe the creature as emitting low growls and a distinctive odor, adding to its fearsome reputation.

Origin

The Nakula legend is rooted in the oral traditions of Alaska's Indigenous communities, who speak of this creature as a spirit of the land, protecting sacred areas and natural resources from over-exploitation. Stories passed down through generations suggest that the Nakula has been present in Alaska's wilderness for centuries, guarding remote parts of the region and instilling respect for the natural world. Many locals regard the Nakula as a symbol of the wild, embodying the strength, resilience, and mystery of the Alaskan landscape.

Lore

In Alaskan lore, the Nakula is often portrayed as a guardian figure, fiercely protective of its territory. Tales describe it as patrolling the vast, untouched forests and mountain ranges, watching over the land from the shadows. Some accounts suggest that the Nakula is highly intelligent and can sense when humans disrespect the land or attempt to harm its habitat. According to local stories, the creature is known to chase away hunters, trappers, and loggers who enter its domain with ill intentions. While encounters are rare, those who have seen the Nakula describe feeling an intense, primal fear, as if sensing its ancient power and connection to the wild.

Additional Facts
- The Nakula is sometimes compared to Bigfoot or the Alaskan Tlingit's Kushtaka, though it is regarded as a distinct entity with its own unique characteristics.

- Legends suggest that the Nakula can appear during times of environmental disruption, serving as a warning to humans to respect the land and its resources.

- In local Indigenous art and storytelling, the Nakula is often depicted as a powerful, wise figure, symbolizing the untamed beauty of Alaska.

Additional Resources

1. The book Alaska's Monster Legends by Donald G. Lehman explores cryptids of the Alaskan wilderness, including the Nakula and its place in local folklore.

69

NANDI BEAR

(Kenya)

Description

The Nandi Bear is a cryptid from East African folklore, particularly in the highlands of Kenya, where it has been feared and revered for generations. Often described as a large, ferocious beast, the Nandi Bear resembles a blend of a hyena and a bear, with thick, dark fur, a sloping back, and muscular, powerful limbs. Witnesses frequently report a massive head with prominent teeth, clawed paws, and a hunched, menacing posture. It is said to stand up to six feet tall on all fours and has a distinctive, bone-chilling roar. Known to be nocturnal, the Nandi Bear is believed to hunt animals, and some accounts suggest it may even pose a threat to humans.

Origin

The legend of the Nandi Bear traces back to the stories of the Nandi people in Kenya, who believed this creature to be a formidable spirit or supernatural being of the wilderness. Known as Kerit in their language, the creature was respected and feared, with stories of its presence dating back centuries. As European explorers and settlers came to East Africa in the late 19th and early 20th centuries, they, too, reported sightings and encounters with the beast. Despite extensive search efforts, no concrete evidence of the creature has ever been found, leaving it an enduring part of African folklore.

Lore

In Kenyan folklore, the Nandi Bear is often portrayed as a mysterious guardian of the forests, only appearing when its territory is threatened or to strike fear in those who dare to venture too close to its domain. It is said to have supernatural abilities, including a stealthy approach and a sudden, terrifying speed when it decides to attack. In some tales, the Nandi Bear is associated with tragic or violent incidents, appearing as a harbinger of misfortune. Villagers share stories of its eerie roars echoing through the night, leaving those who hear it feeling a deep sense of dread.

Additional Facts

- The creature is often compared to the extinct chalicothere, an ancient mammal with a similar build, leading some to speculate that it could be a relict population of a prehistoric species.

- In some regions, the Nandi Bear is believed to protect sacred lands, and local customs advise against disturbing its habitat to avoid invoking its wrath.

- Cryptozoologists and wildlife enthusiasts continue to explore Kenya's highlands, hoping to discover whether the Nandi Bear is indeed an undiscovered species or simply a powerful legend.

Additional Resources

1. African Myths and Legends by Kathleen Arnott includes stories about the Nandi Bear and other African cryptids, offering insights into their cultural significance.

70

NGUMA-MONENE

(Congo)

Description

The Nguma-monene is a cryptid from the Congo, often described as a massive, reptilian creature resembling a prehistoric dinosaur or large serpent. Sightings report it as reaching lengths of up to 30 feet, with a scaly, dark green or gray body and a row of spines or ridges running along its back. Its body shape is said to resemble that of a crocodile or snake, yet it is reported to move with a fluid, undulating motion similar to a serpent gliding through water. Some witnesses describe the creature's head as large and elongated, with a wide mouth lined with sharp teeth, giving it a fierce, predatory appearance.

Origin

The legend of Nguma-monene originated among the indigenous people of the Congo Basin, where it is believed to inhabit the rivers and dense jungles of the region. Local folklore holds that the creature is ancient, a possible remnant from prehistoric times, surviving deep in the remote and largely unexplored wilderness. The name "Nguma-monene" translates to "large python," although it is generally considered to be far larger and more menacing than any known snake species in the area. The cryptid gained some attention in the cryptozoological community in the late 20th century, with explorers and locals recounting sightings in the region.

Lore

Nguma-monene is regarded as a formidable and mysterious entity in local lore, with many stories highlighting its elusive and dangerous nature. The creature is often said to guard its territory fiercely, particularly around the water sources it inhabits. Some legends describe it as a spirit or guardian of the jungle, one that emerges only when provoked or when its habitat is threatened. Stories of the creature's sudden appearances in riverbeds or lakes, often leaving behind large, snake-like tracks, have fueled speculation and fear among those who live near its supposed habitat. It is also said to emit low, rumbling sounds that echo through the jungle, adding to its mystique and fearsome reputation.

Additional Facts
- The creature is often compared to other river or lake cryptids, such as Mokele-mbembe, although Nguma-monene is considered more serpentine in form and behavior.
- Reports of Nguma-monene leave behind large tracks that resemble snake-like imprints, often found near riverbanks or muddy lake shores.
- Its legend has inspired various expeditions by cryptozoologists and adventurers, though no conclusive evidence of its existence has yet been discovered.

Additional Resources

1. Living Dinosaurs? The Search for Prehistoric Relics by Roy Mackal explores the possibility of surviving dinosaurs and cryptids.

71

NINKINANKA

(Gambia)

Description

The Ninki-Nanka is a cryptid from West African folklore, particularly associated with the rivers and swamps of the Gambia. This creature is described as a large, dragon-like being with reptilian features and a lengthy, snake-like body that can measure over 30 feet long. The Ninki-Nanka is often depicted with a scaly hide, a long, winding tail, and a crest or fin running down its back. Some accounts suggest that its head resembles that of a horse or crocodile, with sharp, reflective eyes and a powerful jaw filled with sharp teeth. Known for its elusive and reclusive nature, the Ninki-Nanka is said to be highly territorial, dwelling in secluded areas, particularly deep swamps and remote riverbanks, where it is rarely seen by locals.

Origin

The legend of the Ninki-Nanka dates back centuries and is deeply embedded in the oral traditions of West African tribes. Stories of the creature are most prominent in the Gambia, though similar legends exist across other regions of West Africa. In these tales, the Ninki-Nanka is believed to be a guardian of the waters, a powerful spirit creature that protects its territory and punishes those who dare to intrude. The creature's mystique and fearsome reputation have led to tales warning locals and travelers against venturing too far into uncharted swamp areas, where they risk disturbing the Ninki-Nanka and inviting misfortune upon themselves.

Lore

The Ninki-Nanka is a feared and respected creature in Gambian folklore, and encountering it is believed to bring dire consequences. Local tales recount that those who lay eyes on the Ninki-Nanka often fall ill or die shortly after the encounter, adding to its reputation as a cryptid of ill-omen. Some legends even claim that the creature has the ability to curse those who disturb it. Despite its fearsome reputation, the Ninki-Nanka is also seen as a symbol of mystery and power, an embodiment of the unseen forces that dwell in the natural world. Traditional healers and shamans often invoke the creature in stories to warn of the dangers of disrespecting the natural world.

Additional Facts
- The Ninki-Nanka legend shares similarities with other cryptids across Africa, such as the Kongamato of Zambia, suggesting a regional mythological tradition of dragon-like beings associated with water.

- In Gambian culture, respect for the Ninki-Nanka and similar creatures is integral to local beliefs about maintaining balance with nature.

Additional Resources

1. The Cryptozoology Journal of Africa contains articles on the Ninki-Nanka and its connections to other African cryptids.

2. Myths and Legends of West Africa by Robert Brown offers insight into the folklore of the Gambia, including stories of creatures like the Ninki-Nanka.

72

NØKKEN

(Scandinavia)

Description

The Nøkken is a cryptid-like creature from Scandinavian folklore, often described as a sinister water spirit inhabiting lakes, rivers, and ponds throughout Norway, Sweden, and Denmark. Known for its shape-shifting abilities, the Nøkken can take on various forms, including that of a handsome man, an ethereal figure, or even a monstrous creature with elongated limbs and dark, eerie eyes. In some tales, the Nøkken appears as a beautiful, mesmerizing horse to lure unsuspecting travelers. Typically, it is depicted with dark, algae-covered skin and hair that seems almost part of the water itself, giving it an elusive and spectral appearance. The creature is often accompanied by the faint, haunting sounds of a violin, adding an unsettling element to its presence.

Origin

The Nøkken's origins are rooted in ancient Scandinavian folklore, with its legend passed down through generations as a cautionary tale. Traditionally, the Nøkken was thought to be a guardian or embodiment of the water, representing both its life-giving and treacherous aspects. Myths about water spirits like the Nøkken were particularly common in rural areas where bodies of water were essential to survival, but also dangerous. The Nøkken is often compared to other mythical beings in European folklore, such as the Germanic "Nix" or the Scottish "Kelpie," reflecting a shared fear of the unknown depths. In some regions, it was believed that the Nøkken could only be pacified by casting iron or blood into the water, ensuring safe passage.

Lore

In folklore, the Nøkken is infamous for luring people, particularly children, to the water's edge with enchanting music or shapeshifting abilities. Many stories caution against traveling alone near water at night, warning that the Nøkken might drag unwary wanderers into the depths. The Nøkken's shape-shifting nature allowed it to adapt its appearance to match its victim's desires, often taking the form of a loved one, a beautiful maiden, or a friendly animal. In its true form, the Nøkken is portrayed as having a twisted, eerie appearance, with long limbs and haunting eyes. Legends suggest that anyone who hears its music or sees its form is at risk, as it uses hypnosis to entice its prey.

Additional Facts
- In some Scandinavian villages, protective measures such as casting iron into the water were used to ward off the Nøkken.
- The Nøkken is sometimes portrayed in children's tales as a frightening figure to dissuade them from playing near dangerous waters.

Additional Resources

1. Scandinavian Folk Belief and Legend by Kvideland and Sehmsdorf contains traditional tales, including stories of the Nøkken and other Scandinavian spirits.

73

NUE

(JAPAN)

Description

The Nue is a legendary creature from Japanese folklore, often described as a sinister chimera with an unsettling combination of animal features. Traditionally, it is said to have the head of a monkey, the body of a tiger, the legs of a tanuki (Japanese raccoon dog), and a snake as its tail. Known for its eerie calls and shapeshifting abilities, the Nue often appears at night or during storms, filling the air with an ominous, otherworldly atmosphere. Its unusual appearance adds to its reputation as a symbol of misfortune and chaos.

Origin

The Nue legend dates back to Japan's Heian period (794–1185 AD), a time deeply rooted in supernatural beliefs and spirituality. The most famous story of the Nue appears in The Tale of the Heike, a classic Japanese epic recounting historical events, where a Nue terrorizes the imperial palace, causing illness and misfortune for Emperor Konoe. In the tale, the legendary samurai Minamoto no Yorimasa defeats the creature by shooting it down with an arrow, thus saving the emperor.

Lore

The Nue is often portrayed as a harbinger of bad luck and a bringer of chaos. Folklore suggests that its appearance foretells approaching disaster or misfortune. Known for its chilling screech and mysterious nature, it is said to haunt misty forests and remote mountains, striking fear into those who hear its cries. Due to its association with misfortune, the Nue has become a symbol of Japan's darker supernatural lore, inspiring stories, artwork, and traditional Japanese theater, where it is depicted as a fearsome yet captivating figure.

Additional Facts

- After slaying the Nue, Minamoto no Yorimasa became celebrated for his bravery, and the creature's body was supposedly buried in a remote mountain, further adding mystery to its legend.

- The Nue may reflect ancient Japanese fears of strange and unknown animals, symbolizing humanity's fascination with and fear of the unknown.

- Though primarily mythological, the Nue legend is occasionally evoked in connection with strange animal sightings in rural Japan.

Additional Resources

1. Japanese Tales of Mystery and Imagination by Lafcadio Hearn includes stories and interpretations of yokai, including creatures like the Nue.

2. The National Museum of Japanese History sometimes features exhibits on yokai, showcasing the cultural impact and artistic representations of mythical creatures like the Nue.

74

OGOPOGO

(Canada)

Description

The Ogopogo is a lake monster and cryptid said to inhabit the depths of Okanagan Lake in British Columbia, Canada. Often described as a massive, serpentine creature, the Ogopogo is typically portrayed with a long, undulating body, dark, smooth skin, and multiple humps that rise above the water's surface when it moves. Witnesses often compare its head to that of a horse or a snake, with some accounts describing a pair of large, expressive eyes. Estimated to reach lengths of up to 50 feet, the creature has gained a reputation as both a mysterious and somewhat elusive presence in the lake, known for gliding gracefully along the water.

Origin

The legend of Ogopogo dates back centuries to the Indigenous peoples of the region, who referred to the creature as Naitaka, or "Water Demon." The Indigenous lore portrays it as a powerful spirit associated with the lake, believed to guard the waters and to punish those who trespass without respect. According to oral tradition, sacrifices of small animals were offered to appease the spirit before crossing certain parts of the lake. In the early 20th century, European settlers began reporting sightings of a mysterious creature in Okanagan Lake, and the legend of Ogopogo gained widespread attention.

Lore

The Ogopogo is often viewed as a benevolent yet intimidating guardian of Okanagan Lake, embodying the mysteries and dangers of nature. In local folklore, the creature is thought to reside in the lake's darkest depths, emerging only on rare occasions or when disturbed. Sightings often describe the creature as moving with an undulating, snake-like motion, and it is known to surface during calm weather or at dusk. Although most encounters are brief and non-threatening, there is a longstanding belief that the creature's presence serves as a reminder of the lake's ancient, untamed power. Ogopogo's legend has inspired numerous expeditions, folklore, and speculation, with tourists and researchers drawn to the lake each year in hopes of witnessing the elusive beast.

Additional Facts
- In recent years, sonar technology and underwater cameras have been used to search for Ogopogo, though definitive evidence remains elusive.
- The creature is celebrated in Okanagan culture, with statues, murals, and local festivals dedicated to its legend, contributing to the region's tourism and folklore.

Additional Resources

1. The Okanagan Valley Museum offers historical and cultural insights into Ogopogo's legend, with exhibits on local folklore.

2. Monsters of the Deep: Exploring North America's Lake Monsters by Thomas Slemen delves into Ogopogo and other aquatic cryptids, exploring theories and witness accounts.

75

OKONKORO

(Cameroon)

Description

The Okonkoro is a cryptid from the dense forests and remote regions of Cameroon, described as a large, ape-like creature with a mysterious and intimidating presence. Reported sightings depict it as standing between six and eight feet tall, with broad shoulders, a muscular build, and covered in thick, dark fur that helps it blend seamlessly with the shadows of the rainforest. The Okonkoro's face is often described as a mix between that of a gorilla and a bear, with intense, deep-set eyes that appear to watch intently before vanishing into the dense jungle. Witnesses sometimes report loud, echoing vocalizations attributed to the creature, adding to its elusive and unsettling reputation.

Origin

The Okonkoro legend is rooted in the folklore of the indigenous people of Cameroon, where stories of this powerful creature have been passed down through generations. It is often associated with the untouched and mysterious regions of the rainforest, particularly near the borders of Cameroon, Gabon, and the Central African Republic. Known to the local population as a guardian of the wilderness, the Okonkoro is thought to protect the forests from intruders and may emerge only when its territory is threatened. Its existence has also attracted interest from cryptozoologists and adventurers, though no concrete evidence has yet confirmed its presence.

Lore

In local lore, the Okonkoro is considered a reclusive spirit of the forest, rarely encountered by humans and viewed with both reverence and caution. Legends describe it as a creature that prefers solitude, avoiding human contact while fiercely guarding its domain. Some stories suggest that the Okonkoro possesses supernatural abilities, such as the power to move silently or to blend seamlessly into the forest, making it nearly impossible to track. Those who claim to have encountered the Okonkoro speak of an intense feeling of being watched, followed by a sudden silence, as if the creature is studying them before retreating back into the depths of the jungle.

Additional Facts

- The Okonkoro is often associated with unexplained animal deaths in the area, as locals believe the creature sometimes hunts large prey at night.

- Its deep, resonant vocalizations, reportedly heard echoing through the forest, have led some to speculate that the Okonkoro could communicate with other creatures or with members of its kind.

- In Cameroonian culture, the Okonkoro is viewed as a symbol of the wild, untouched forest, embodying both the beauty and mystery of the natural world.

Additional Resources

1. *Cryptids and Monsters of Africa* by Richard Freeman explores lesser-known African cryptids, including the Okonkoro and its cultural significance in Cameroon.

OWEBRE

(Nigeria)

Description

The Owebre is a cryptid from Nigerian folklore, particularly known in the dense forests and remote regions of Nigeria. Described as an elusive, mysterious creature, the Owebre is said to resemble a large, ape-like figure with some feline characteristics. It is often depicted as having a muscular build, dark fur, and a tail similar to that of a lion, blending traits of both predator and primate. Witnesses describe it as standing on two legs, towering at around six to eight feet tall, with piercing eyes that seem to glow faintly in the dark forest. Some accounts mention that its face combines both ape and cat-like features, lending it an eerie, almost mythical appearance.

Origin

The origins of the Owebre legend can be traced to indigenous Nigerian tribes who live close to thick jungles and remote forest regions. Many believe the creature has existed for generations, with stories passed down through local communities. The Owebre is often seen as both a guardian and a threat, its presence signifying both respect for nature and caution about the dangers of the wilderness. While some think the Owebre may be a mythological relic symbolizing untamed wildlife, others consider it a genuine cryptid, hiding within the dense, uncharted forests of Nigeria.

Lore

The Owebre is both feared and revered by local villagers. According to lore, the creature possesses remarkable agility and strength, able to move swiftly and silently through the thick underbrush. Some legends describe it as a protector of the forest, deterring intruders and hunters who exploit the land. Others tell of frightening encounters, where the Owebre is said to let out a loud, bone-chilling roar, driving people away from its territory. Sightings are rare, and those who claim to have seen the Owebre are often reluctant to revisit the same area, as the creature is believed to mark those who trespass its domain.

Additional Facts
- Local hunters and villagers have reported finding large, clawed footprints that resemble a blend of feline and primate prints, adding to the mystery of the creature.
- Some believe that the Owebre has a mystical connection to the forest spirits, guarding certain regions from human exploitation.
- The Owebre is rarely seen close to settlements, and it is most often associated with deeply forested and mountainous areas where few humans venture.

Additional Resources

1. The Mystery of Nigerian Cryptids: An Exploration of Local Legends by Chinedu Eze provides insight into the country's lesser-known creatures, including the Owebre.

77

PAKTANAK

(PHILIPPINES)

Description

The Paktanak is a malevolent entity in Philippine folklore, often portrayed as a vengeful female spirit driven by anguish and hatred. It takes the form of a gaunt woman with tangled, unkempt hair, piercing bloodshot eyes, and decaying skin. Some accounts describe it with unnaturally elongated limbs and razor-sharp claws, adding to its terrifying visage. Known for its haunting, bone-chilling wail, the Paktanak's cry serves as an omen of its approach. It is said to haunt remote locations like dense forests or abandoned houses, targeting those who have violated sacred spaces or betrayed others.

Origin

The Paktanak originates from the folklore of rural Philippine communities, particularly in the Visayas and Mindanao regions. Believed to be the restless spirit of a woman who met a tragic or violent end, the Paktanak embodies the consequences of betrayal, cruelty, or unfulfilled desires. Its legend, passed down through oral traditions, serves as both a cautionary tale and a moral warning about the repercussions of wrongdoing and the importance of respect for sacred boundaries.

Lore

Folklore describes the Paktanak as a spirit consumed by sorrow and rage, seeking vengeance on those who disrespect others or fail to honor their promises. One story recounts how it appears as a beautiful woman to lure its victims, only to reveal its monstrous form before striking. Its mournful wail is believed to unnerve its prey, driving them into madness before it attacks. In some traditions, it is said that the Paktanak can be appeased through rituals, prayers, or offerings, helping it find peace and move on to the afterlife. Additionally, stories of its wrath often serve as warnings to respect women and honor the sanctity of agreements.

Additional Facts
- It is believed that wearing amulets or charms, such as the anting-anting, can offer protection from its wrath.
- In some communities, candles and food are left as offerings to prevent its appearance or appease its spirit.

Additional Resources

1. Maximo D. Ramos (1971), The Aswang Complex in Philippine Folklore

2. Francisco R. Demetrio, Philippine Myths, Legends, and Folktales

3. Daminon Journal, Spirits of Vengeance: A Study on Philippine Ghosts and Folklore

78

PESTIMAI

(INDONESIA)

Description

The Pestimai is a fearsome and mysterious entity from Indonesian folklore, often portrayed as a nocturnal creature with an emaciated, human-like form. It has glowing red eyes, elongated limbs, and leathery bat-like wings, allowing it to glide silently through the night. Its pale, sickly skin appears cracked or mottled, with sharp claws at the tips of its bony fingers. Known for exuding an aura of dread, the Pestimai's raspy, guttural breathing serves as an ominous warning of its approach. It preys on defenseless victims, often targeting those who are sleeping or isolated during the dark hours.

Origin

The Pestimai's legend originates from rural regions of Indonesia, where it is believed to be a malevolent spirit or a cursed soul condemned to eternal hunger. Local folklore suggests it was once human, transformed by acts of betrayal, greed, or other grave moral failings. Associated with dark, abandoned places like caves, dense jungles, and derelict homes, the Pestimai uses these locations as its lairs, lurking in the shadows to strike.

Lore

In Indonesian lore, the Pestimai feeds on the life force of its victims, leaving them frail or gravely ill. It is especially feared for its attacks on newborns and pregnant women, likening it to similar vampiric or demonic entities found in Southeast Asian folklore. Villagers often recount hearing eerie breathing or the sound of wings flapping in the dead of night, incidents frequently accompanied by the unexplained illness or death of livestock. Protective measures include hanging garlic or chili peppers outside homes, lighting candles, and conducting shaman-led rituals to drive the Pestimai away.

Additional Facts
- It is said to be unable to enter homes without an invitation, making locked doors and windows essential.
- Salt or holy water is believed to weaken or repel the Pestimai.
- Some versions of the tale suggest that the Pestimai's curse can be lifted through specific rituals or acts of redemption.

Additional Resources

1. J. F. A. Swellengrebel, Myth and Reality in Indonesian Folklore
2. Clara van der Laan, Demonic Spirits in Southeast Asian Cultures
3. Dipa Nusantara Aidit, Folktales of Java and Sumatra.

79

PONTIANAK

(Malaysia)

Description

The Pontianak is a fearsome figure in Malaysian folklore, believed to be the vengeful spirit of a woman who died tragically during childbirth. Often described as pale-skinned with long, flowing black hair and piercing red eyes, the Pontianak blends beauty with terror to lure unsuspecting victims. She haunts dark, secluded areas, often near banana trees, and her high-pitched, chilling laughter warns of her presence. As she approaches, the laughter becomes softer, signaling her deadly intent. When revealed in her true form, the Pontianak transforms into a horrific entity with sharp claws and bloodied garments, embodying her rage and sorrow.

Origin

The Pontianak's name is derived from "perempuan mati beranak", meaning "woman who died in childbirth," reflecting her tragic origins. Legends from Malaysian villages recount that the Pontianak arises from unfulfilled vengeance or grief, driven to exact retribution on men or anyone who crosses her path. Her existence is deeply rooted in traditional beliefs surrounding the restless dead, and her story serves as a cautionary tale about the consequences of betrayal and violence against women.

Lore

The Pontianak is infamous for preying on men, often using her beauty to captivate them before revealing her monstrous form. She is said to feed on internal organs, particularly hearts, leaving her victims mutilated or lifeless. Her presence is marked by the fragrance of frangipani flowers or a rotting stench, depending on her mood. Banana trees are believed to be her preferred resting place, and villagers often tie red threads around these trees to trap her. Rituals, prayers, or sharp objects like needles and nails are used as defenses against her wrath.

Additional Facts
- The Pontianak is closely associated with the Langsuir, another female ghost in Southeast Asian mythology, though the two have distinct stories and traits.

- Her laughter is a warning: soft laughter signals her proximity, while loud laughter means she is farther away.

- Sharp objects such as needles or nails are believed to weaken her; inserting them into the back of her neck can temporarily subdue her.

- Folklore surrounding the Pontianak is often used to instill respect for women and caution against neglect or abuse.

Additional Resources

1. E. Koentjaraningrat, Traditional Ghosts of Southeast Asia.

80

POPOBAWA

POPOBAWA

Description

The Popobawa is a fearsome and infamous cryptid from the folklore of Zanzibar and the surrounding East African islands. This malevolent, shape-shifting entity is often described as a bat-like humanoid with leathery wings, sharp claws, a single glowing eye, and a pointed tail. Its name, meaning "bat-wing," reflects its terrifying appearance. The Popobawa is notorious for its nocturnal visits, during which it terrorizes individuals or entire households. Known for spreading fear and psychological torment, the creature leaves its victims shaken and often ostracized by their communities.

Origin

The Popobawa legend emerged in the mid-20th century, closely tied to the social and political upheavals following the Zanzibar Revolution of 1964. Folklorists theorize that the cryptid represents the collective anxieties of a community grappling with rapid change and instability. Others connect the Popobawa to Islamic beliefs in malevolent djinn or spirits, suggesting it may have older roots in the region's spiritual traditions.

Lore

Stories of the Popobawa describe it as erratic and terrifying, often targeting multiple victims in a single night. It is said to attack its prey while they sleep, immobilizing them with its presence. Many victims describe waking to find the creature pressing down on their chest, rendering them unable to move or scream. The Popobawa is believed to draw power from fear, thriving on the terror it inspires. According to folklore, openly acknowledging or discussing its presence can repel the creature, while silence invites further attacks. Protective measures include burning incense, reciting Quranic verses, and seeking help from traditional healers.

Additional Facts

- The Popobawa is said to shift between physical and spectral forms, making it elusive and difficult to confront.
- It is often associated with periods of social unrest or election seasons, when fear and tension in the community are heightened.
- Some experts believe the Popobawa legend serves as a cautionary tale to enforce social order and discourage disobedience.
- Signs of the Popobawa's presence include foul odors, sudden chills, and ominous shadows.

Additional Resources

1. G. M. Tilyard, Spirits and Shadows: Folklore of East Africa.
2. Abdulaziz Lodhi, Tales of Zanzibar: Myths and Urban Legends.

81

RAKE

(USA)

Description

The Rake is a chilling cryptid from modern American folklore, described as a pale, hairless humanoid with grotesque features. Its elongated limbs, sharp claws, and glowing eyes give it a predatory and otherworldly appearance. Emaciated and hunched, the Rake moves silently, lurking in forested areas or creeping into homes under the cover of darkness. Victims often report an intense, inexplicable dread accompanying its presence, as though the creature exudes an unnatural malevolence.

Origin

The Rake originated as a fictional character in internet creepypasta stories, first gaining attention in the early 2000s. These short, user-generated horror stories were shared across forums such as 4chan and the Creepypasta Wiki, where the creature's eerie mythology began to take shape. Despite its fictional origins, the Rake has transcended its internet roots, becoming a staple of contemporary folklore. Its name comes from the sharp, rake-like claws it is said to use when attacking its victims.

Lore

The Rake is often portrayed as a patient and predatory entity, stalking its victims before making its move. Stories describe it entering homes at night, standing ominously at the foot of beds as it watches its prey. Some accounts suggest the Rake communicates telepathically, implanting horrifying thoughts or visions into the minds of its victims. Those who survive encounters frequently report lingering effects such as insomnia, paranoia, and vivid nightmares. In many tales, the Rake's presence is tied to remote forests or desolate areas, symbolizing the hidden dangers of the unknown.

Additional Facts
- The Rake is frequently compared to cryptids like the Wendigo or Skinwalker due to its humanoid form and predatory habits.
- Encounters often involve a chilling silence or a sudden drop in temperature, both considered signs of its presence.
- Though created as a fictional character, the Rake's widespread popularity has blurred the line between myth and reality.
- Its story has inspired countless videos, artwork, and horror media, cementing its place as a modern icon of terror.

Additional Resources

1. Michael Newton, Encyclopedia of Cryptozoology.

2. Dario Nardi, Modern Myths and Urban Legends.

3. Creepypasta Archives: Stories of the Rake on Creepypasta Wiki and Reddit's /r/NoSleep.

82

RASPUTIN'S DRAGON

(Russia)

Description

Rasputin's Dragon is a cryptid from Russian folklore, intricately tied to the enigmatic figure of Grigori Rasputin. It is described as a massive, serpentine beast with black, shimmering scales that glimmer with an iridescent sheen, resembling oil on water. Its glowing red eyes burn with an otherworldly intensity, while its ragged, membranous wings add to its fearsome presence. Legends suggest that the creature was either a guardian summoned by Rasputin's mystical abilities or a manifestation of the dark powers he was believed to possess. Silent and stealthy, Rasputin's Dragon is said to strike suddenly, instilling terror in those who encounter it.

Origin

The legend of Rasputin's Dragon is rooted in the historical mystique surrounding Grigori Rasputin, whose reputation as a healer and sorcerer inspired countless tales. Combining elements of Siberian folklore with Rasputin's notoriety, the story describes a shadowy dragon-like entity tied to him. Some believe the creature was an ancient guardian spirit bound to Rasputin through arcane rituals, while others suggest it was a curse inflicted upon him due to his alleged dabbling in forbidden powers. The dragon's origins remain a topic of speculation, with its mythology evolving through generations of storytelling.

Lore

Folklore surrounding Rasputin's Dragon paints it as a fearsome protector, emerging from the darkness to shield Rasputin during moments of danger. It is said to emit a low, bone-chilling growl that paralyzes its enemies with fear. According to some accounts, the dragon played a role in Rasputin's infamous resilience, helping him survive multiple assassination attempts. After his death, the dragon reportedly vanished, though sporadic sightings have been reported in the remote forests of Siberia and the Ural Mountains. Some legends claim the dragon still guards hidden relics associated with Rasputin's mystical knowledge.

Additional Facts

- Rasputin's Dragon is said to be nearly invisible in low light, earning it the nickname "Shadow Serpent."
- Its cold, misty breath is believed to freeze water instantly and chill the air around it.
- Encounters with the dragon are often associated with an overwhelming sense of dread and an instinct to flee.
- Some myths suggest that the dragon guards a secret cache of relics tied to Rasputin's alleged occult practices.

Additional Resources

1. Alexander Spiridovich, Rasputin: The Myth and the Man.

2. Russian Folklore Archives, Legends of the Shadow Serpent.

ROPEN

(Papua New Guinea)

Description

The Ropen is a cryptid from the folklore of Papua New Guinea, often described as a large, winged creature bearing a striking resemblance to a prehistoric pterosaur. It is characterized by leathery wings, a long tail tipped with a diamond-shaped flange, and a bioluminescent glow that lights up the night sky. With a reported wingspan of 10 to 30 feet, the Ropen's glowing appearance lends it an aura of mystery and otherworldliness. Witnesses often describe its piercing screeches as it glides over coastal regions and remote islands, adding to its eerie reputation.

Origin

The Ropen legend is deeply embedded in the oral traditions of the indigenous peoples of Papua New Guinea, especially those living in isolated coastal areas and islands. The term "ropen," in certain local dialects, translates to "demon flyer," reflecting its feared and supernatural status. While the creature has roots in mythology, modern sightings have sparked interest among cryptozoologists, with theories ranging from it being a surviving pterosaur to an unidentified species of bat or bird.

Lore

Local folklore portrays the Ropen as a nocturnal and ominous figure, both feared and respected by the communities it visits. In some traditions, its appearance near villages foretells calamity, earning it the label of a harbinger of doom. Others describe it as a scavenger, feeding on fish by the shore or even raiding burial sites for human corpses. Sightings are most often reported near coastal caves or deep, forested areas, thought to be its daytime resting places. Its bioluminescent glow, often described as a rhythmic pulsing light, is seen as either a natural adaptation or a supernatural trait.

Additional Facts

- The Ropen's glowing bioluminescence has been likened to the light of fireflies or deep-sea creatures, leading to speculation about a chemical basis for the phenomenon.

- It is sometimes compared to Indonesia's Ahool, a similar bat-like cryptid, though the Ropen's pterosaur-like features set it apart.

- Expeditions to Papua New Guinea have resulted in numerous anecdotal reports and blurry photographs, but conclusive evidence remains elusive.

- Skeptics suggest that sightings of the Ropen could be misidentifications of frigate birds, large bats, or other known animals.

Additional Resources

1. Jonathan David Whitcomb, Searching for Ropens: Living Pterosaurs in Papua New Guinea.

84

SCORCHING DEATH

(Brazil)

Description

The Scorching Death is a cryptid rooted in Brazilian folklore, described as a blazing humanoid figure that haunts the depths of the Amazon rainforest. Engulfed in flames, the creature has glowing red eyes that pierce the darkness, and some accounts depict it with clawed hands and a molten, skeletal frame. Legends state that it radiates an unbearable heat, igniting vegetation and severely burning anyone who dares approach. Witnesses often describe hearing crackling, eerie laughter accompanying the creature's fiery presence, adding to its terrifying reputation.

Origin

The Scorching Death originates from indigenous myths of the Amazon, often tied to spirits of vengeance or guardianship. Some traditions depict it as the restless spirit of someone who perished in a fire, cursed to wander and bring destruction. Other accounts suggest it is a supernatural guardian, protecting sacred lands from exploitation and punishing those who harm the rainforest. These tales likely evolved as cautionary stories to promote respect for the natural environment and its dangers, including the risk of wildfires.

Lore

In local lore, the Scorching Death is both feared and revered, acting as a harbinger of destruction and a protector of nature's balance. The cryptid is said to appear in areas affected by deforestation or environmental harm, emerging from the shadows to mete out punishment. Witnesses often recount an overwhelming wave of heat, followed by the sight of the fiery figure advancing through the jungle. It is believed to leave scorched vegetation and footprints in its wake, serving as a grim reminder of its wrath. Indigenous communities sometimes perform rituals or offerings to appease the spirit, seeking forgiveness or protection.

Additional Facts

- The Scorching Death is often compared to other fire-associated myths, such as the Will-o'-the-Wisp or the Brazilian Boitatá, though it is far more menacing and destructive.

- Its presence is often accompanied by a sudden rise in temperature, smoke, and the acrid smell of burning wood or sulfur.

- Skeptics suggest sightings could be explained by natural phenomena like lightning strikes, forest fires, or even hallucinations caused by extreme heat or stress in the jungle.

- Despite its fearsome image, some locals view the Scorching Death as a necessary guardian that enforces respect for the rainforest and its spiritual significance.

Additional Resources

1. Luis da Silva, Myths of the Amazon: Spirits and Legends.

85

SINKHOLE SAM

(Kansas, USA)

Description

Sinkhole Sam is a cryptid from Kansas folklore, rumored to inhabit the murky waters of large sinkholes near Inman. Described as a massive, eel-like creature, Sam is said to measure between 15 and 30 feet in length, with a slick, cylindrical body covered in dark gray or greenish skin. Its wide, flat head features large, lidless eyes and a gaping, toothless mouth, lending it an eerie and otherworldly appearance. Witnesses report that Sinkhole Sam moves with remarkable speed and agility, making it elusive and difficult to observe for long. The creature is believed to survive on fish and other small aquatic animals.

Origin

The story of Sinkhole Sam emerged during the mid-20th century, sparked by reports of a strange, unidentified creature in Carey Lake, a large sinkhole formed during the Dust Bowl era. Some locals speculated that Sam might be an ancient species trapped in subterranean aquifers beneath the region. Folklorists have drawn connections between Sinkhole Sam and traditional myths of lake monsters or prehistoric creatures, suggesting the tale reflects both modern curiosity and historical storytelling traditions.

Lore

According to local lore, Sinkhole Sam is a shy but formidable creature, avoiding humans and rarely surfacing. Sightings often describe it briefly emerging to bask or investigate boats, only to quickly retreat into the water when approached. Many believe the sinkholes and their vast, interconnected waterways provide a hidden sanctuary for the creature. While most accounts suggest Sam is harmless, some warn that it could capsize small boats or pose a danger to unwary swimmers. Its elusive nature has contributed to its mystique and made it a fascinating subject of local folklore.

Additional Facts
- The creature is often compared to other aquatic cryptids like the Loch Ness Monster due to its elongated, serpentine shape.
- Skeptics argue that sightings may stem from misidentifications of large catfish, freshwater eels, or even drifting logs.
- Inman has embraced the legend as part of its cultural identity, hosting occasional searches and discussions about the mysterious inhabitant of Carey Lake.

Additional Resources

1. R.L. Duckworth, Hidden Creatures of the American Heartland.
2. Kansas Folklore Society, Tales from the Plains: Myths and Legends of Kansas.
3. F. Matthews, Cryptids of Middle America: A Regional Guide.

86

SKINWALKER

(Southwest USA)

Description

The Skinwalker is a terrifying entity from Navajo folklore, described as a witch or shape-shifter capable of transforming into animals or mimicking human forms. In its human guise, it appears gaunt and unnerving, with glowing eyes that pierce the darkness. When transformed, the Skinwalker often takes the shape of predatory animals such as wolves, coyotes, or owls, though some accounts describe it as a grotesque fusion of human and animal traits. Renowned for its supernatural speed, agility, and ability to mimic human voices, the Skinwalker uses these powers to manipulate, terrify, and hunt its victims.

Origin

The Skinwalker, or yee naaldlooshii in Navajo, meaning "by means of it, it goes on all fours," is deeply rooted in Navajo tradition. According to legend, Skinwalkers are malevolent witches who gain their powers through heinous acts, such as murdering a close family member or breaking significant cultural taboos. By abandoning their humanity, they acquire dark abilities that set them apart from other beings. These stories often serve as cautionary tales within Navajo communities, emphasizing the importance of moral conduct and respect for tradition.

Lore

Skinwalkers are said to roam the remote deserts and mesas of the American Southwest, haunting desolate areas and preying on the unwary. They are frequently blamed for unexplained events, including mysterious deaths, illnesses, and animal attacks. Witnesses describe them as cunning and highly dangerous, able to evade pursuit and outwit even the most seasoned trackers. The creature's ability to mimic voices, often imitating loved ones or cries for help, is one of its most feared traits. Skinwalkers are believed to curse victims by meeting their gaze, while traditional Navajo rituals or sacred items like ash or turquoise are said to offer protection.

Additional Facts
- Their ability to mimic voices is often described as disturbingly accurate, used to lure victims into isolation.
- According to lore, the only way to kill a Skinwalker is by uncovering its true identity and exposing it publicly, often through a specific ritual.
- Many Navajo people refrain from discussing Skinwalkers openly, as it is believed to attract their attention or summon their presence.

Additional Resources

1. Tony Hillerman, The Blessing Way (fiction incorporating Navajo culture and Skinwalker themes)
2. David Weatherly, Strange Intruders.

87

SKUNK APE

(Florida, USA)

Description

The Skunk Ape, also called the Florida Bigfoot, is a cryptid from the folklore of the southeastern United States, particularly associated with Florida's Everglades. It is described as a large, ape-like creature, standing between 6 and 8 feet tall and covered in dark reddish-brown or black fur. One of its most distinctive traits is its overpowering odor, often compared to a mix of decaying vegetation and wet animal fur. Witnesses frequently report its glowing, reflective eyes at night, enhancing its eerie reputation. The Skunk Ape is believed to be shy and elusive, navigating swampy terrain with ease and moving silently through dense vegetation.

Origin

The legend of the Skunk Ape traces back to Native American folklore, where stories of a "wild man of the woods" were passed down through generations. In modern times, the legend gained traction in the mid-20th century with reports of sightings emerging from Florida's swamps. Some believe the Skunk Ape is a remnant of an ancient hominid species or a relative of Sasquatch, uniquely adapted to Florida's subtropical environment. Others regard it as a mythological figure, a cautionary tale to respect the dangers of the swamp.

Lore

According to local lore, the Skunk Ape is a nocturnal and solitary creature, avoiding human contact but occasionally raiding campsites or farms for food. Witnesses have reported hearing guttural growls, shrieks, and heavy footsteps breaking through underbrush. The creature is said to have a particular fondness for fruit, especially oranges, which grow abundantly in the region. Sightings often involve brief glimpses of the creature moving through the swamp, leaving broken vegetation and its infamous stench in its wake.

Additional Facts
- The Skunk Ape is considered smaller than Sasquatch but better adapted to swampy environments.
- Its presence has been linked to natural phenomena like swamp gas, which skeptics argue may explain some sightings.
- In 2000, the Skunk Ape Headquarters in Ochopee, Florida, claimed to receive photographs of the creature, reigniting public fascination.
- Although most sightings occur in the Everglades, reports have also surfaced in Georgia, Alabama, and Mississippi.

Additional Resources

1. Loren Coleman, The Field Guide to Bigfoot and Other Mystery Primates.

2. David Shealy, The Skunk Ape Files: Sightings and Evidence in the Florida Everglades.

88

SLENDERMAN

(Worldwide)

Description

Slenderman is a tall, faceless entity that began as an internet urban legend and has become one of the most infamous cryptids in modern folklore. Standing between 8 and 10 feet tall, he is unnaturally thin, dressed in a black suit with a white shirt and tie, resembling a distorted human silhouette. His face is featureless—smooth and blank—lacking eyes, a nose, or a mouth, which amplifies his unsettling appearance. Slenderman is said to have long, spindly arms that extend into tentacle-like appendages, which he uses to stalk, manipulate, or ensnare his victims. Silent and omnipresent, he instills fear and paranoia in those who encounter him, often targeting children.

Origin

Slenderman's origins trace back to 2009 on the Something Awful forums, where users were tasked with creating paranormal images. Eric Knudsen, using the pseudonym Victor Surge, submitted doctored photographs of children with a tall, shadowy figure lurking in the background, accompanied by unsettling captions. These images sparked a surge of user-generated stories, videos, and artwork, elevating Slenderman from an internet meme to a cultural phenomenon. Though fictional, Slenderman has transcended his digital roots, becoming a modern myth deeply ingrained in popular culture and urban folklore.

Lore

In folklore, Slenderman is depicted as a malevolent and enigmatic force that preys on vulnerable individuals, particularly in isolated locations like forests, abandoned buildings, and playgrounds. Witnesses describe an overwhelming sense of dread, nausea, or disorientation when he is near, often accompanied by strange static interference in electronic devices. Slenderman is said to teleport, invade dreams, and manipulate minds, sometimes compelling victims to commit horrifying acts. Some stories claim he abducts his victims to another realm, while others suggest he leaves them wandering aimlessly under his control.

Additional Facts

- Slenderman gained widespread popularity through the web series Marble Hornets, which reimagined him as a sinister, supernatural entity called "The Operator."

- The legend has influenced real-life events, most notably a 2014 incident in Wisconsin involving two teenagers, highlighting the power of digital folklore.

- He is often associated with "Slender Sickness," a fictional condition characterized by paranoia, hallucinations, and memory loss.

Additional Resources

1. Eric Knudsen, Original Post on Something Awful (2009).

89

SNAKE MAN

(India)

Description

The Snake Man is a cryptid from Indian folklore, described as a humanoid figure with striking serpentine features. He is said to have a human-like torso covered in shimmering, scale-like patterns that reflect light, making him appear otherworldly. His eyes are reptilian, with glowing slitted pupils, and his movements are unnervingly fluid and snake-like. In some accounts, the Snake Man possesses a forked tongue and sharp, fang-like teeth capable of delivering venomous bites. Witnesses report that he can extend his limbs like a serpent and slither with unnatural speed, making him both elusive and terrifying. He is often associated with dense forests, ancient temples, and remote villages, where his presence inspires both fear and reverence.

Origin

The legend of the Snake Man is deeply rooted in Indian mythology and rural traditions, bearing similarities to the Nāga, revered serpent deities in Hinduism and Buddhism. Folklore often portrays the Snake Man as a guardian of sacred treasures or cursed relics, tasked with protecting them from desecration. Other stories depict him as a human transformed into a serpent as punishment for betrayal or transgressions. These tales have been passed down for centuries, reflecting themes of divine retribution, justice, and the mystical connection between humans and nature.

Lore

Encounters with the Snake Man are said to occur near water bodies, dense jungles, or ancient ruins. Witnesses often describe him slithering through underbrush or scaling walls with an inhuman grace. Some claim he has a hypnotic gaze, which can entrance his victims or compel them to reveal secrets. Villagers believe encountering the Snake Man brings misfortune unless offerings or rituals are performed to appease him. Protective measures often include wearing talismans or invoking blessings to guard against his influence. Despite his fearsome reputation, some traditions suggest the Snake Man only attacks those who harm sacred sites or the environment.

Additional Facts

- His venom is said to affect both the body and spirit, causing physical ailments or bad luck to those he bites.

- He is believed to be highly intelligent, fiercely protective of his territory, and a symbol of nature's wrath against human exploitation.

Additional Resources

1. Subhash Bhatt, Myths and Creatures of Indian Folklore.

2. Ramesh Menon, Serpents in Indian Mythology: Guardians and Foes.

3. Indian Folklore Journal, Cryptids and Spirits of Rural India.

90

TANIWHA

(New Zealand)

Description

The Taniwha is a legendary creature in Māori folklore, described as a powerful and supernatural being that can appear as either a guardian or a fearsome predator. Often depicted as large, serpentine or dragon-like, Taniwha are said to have scaly, lizard-like bodies, sharp claws, and formidable teeth. Some versions of the legend describe them with fish-like features, such as fins or gills, emphasizing their strong connection to water. They are believed to inhabit rivers, lakes, caves, and oceanic depths, with their presence tied to specific locations or landmarks. While many stories portray Taniwha as protectors of sacred sites, others warn of their capacity for destruction and misfortune.

Origin

The Taniwha holds a prominent place in Māori mythology and oral tradition, revered as a spiritual being deeply connected to the natural world. According to legend, Taniwha were created by the gods to serve as kaitiaki (guardians) of significant areas, such as waterways or tribal lands. In other tales, Taniwha are seen as ancestral spirits or manifestations of nature's raw power. Their dual nature—both protective and destructive—symbolizes their intimate relationship with the environment and underscores the need to respect and honor the land and water.

Lore

Taniwha are often said to guard specific sites, such as hidden caves, underwater passages, or stretches of river, serving as protectors of the people or treasures associated with these areas. In return, local iwi (tribes) offer prayers, food, or other gestures of respect to maintain harmony. However, legends also caution against disturbing their domains, as such actions could provoke a Taniwha's wrath. These stories recount Taniwha unleashing floods, capsizing boats, or attacking those who encroach on their territory. Despite their fearsome reputation, Taniwha are also revered as symbols of strength, deeply intertwined with Māori spiritual and cultural identity.

Additional Facts

- Certain landmarks in New Zealand, such as Te Waikoropupū Springs and Lake Tarawera, are tied to Taniwha legends.
- Some iwi view Taniwha as ancestral beings, while others see them as monsters or cautionary figures warning against environmental destruction.
- The Taniwha remains a significant cultural symbol, inspiring art, literature, and ceremonial practices in modern Māori society.

Additional Resources

1. Margaret Orbell, The Illustrated Encyclopedia of Māori Myth and Legend
2. Sir George Grey, Polynesian Mythology

91

TIKBALANG

(PHILIPPINES)

Description

The Tikbalang is a cryptid from Filipino folklore, often described as a humanoid figure with a horse's head, long limbs, and a muscular body. Towering over humans, its elongated legs give it uncanny speed and agility, allowing it to move effortlessly through dense forests and rugged terrain. Its glowing red eyes and sharp teeth enhance its menacing appearance, and its chilling, otherworldly laughter echoes through the night. While often associated with mischief, the Tikbalang is feared for its ability to confuse and lead travelers astray, leaving them hopelessly lost in the wilderness.

Origin

The Tikbalang originates from pre-colonial Filipino mythology, where it is regarded as a spirit of the wild, deeply connected to nature. It is often seen as a guardian of forests, mountains, and other untamed areas, ensuring that humans respect these sacred spaces. Its mischievous acts are thought to punish those who harm the environment or break cultural norms. Some legends trace the Tikbalang's roots to animistic beliefs, where spirits or deities manifested as animal-human hybrids to interact with people.

Lore

Folklore describes the Tikbalang as a trickster that uses supernatural powers to disorient travelers. It creates illusions, warps the landscape, and alters perceptions of time and space to confuse its victims. Those who encounter the Tikbalang are said to break free from its spell by turning their clothes inside out, a gesture believed to either confuse or appease the creature. Offerings of tobacco or food are also thought to placate the Tikbalang and ensure safe passage through its territory. Despite its fearsome reputation, some stories depict the Tikbalang as a benevolent guardian that protects those who respect the natural world.

Additional Facts
- In some regions, it is said that the Tikbalang can shapeshift into human form to deceive or seduce its victims.
- Legends warn against disturbing its territory, as doing so may provoke its wrath or lead to misfortune.
- Modern depictions of the Tikbalang appear in Philippine literature, art, and media, symbolizing the enduring influence of local folklore.

Additional Resources

1. Francisco Demetrio, Philippine Mythology.

2. Maximo D. Ramos, Creatures of Philippine Lower Mythology.

3. F. Landa Jocano, The Folk Healers of the Philippines: Myth and Reality.

92

TROLL

(Norway)

Description

The Troll is a cryptid originating from Norwegian folklore, described as a large humanoid with grotesque features and an uncanny ability to blend into its rugged environment. Trolls are said to have thick, rock-like skin, elongated limbs, and exaggerated facial features, including large noses and faintly glowing, sunken eyes. Towering between 7 and 15 feet tall, these creatures are reclusive and highly territorial. Witnesses report that trolls move with surprising agility despite their size, and their low, guttural growls echo eerily across the wilderness. Their presence is most commonly associated with remote mountains, dense forests, and hidden caves.

Origin

From a cryptid perspective, trolls are speculated to be an undiscovered species of hominid or animal adapted to Norway's remote and challenging terrain. Unlike their mythological depictions, this view suggests that sightings and stories about trolls may have arisen from rare encounters or misinterpreted natural phenomena, such as large boulders thought to be petrified trolls. Cryptozoologists propose that trolls could be ancient creatures, surviving in isolation far from human activity.

Lore

Trolls are believed to be nocturnal, venturing out only under the cover of darkness. Folklore often describes them as highly sensitive to sunlight, which can disorient them or, in some accounts, turn them to stone—possibly reflecting their preference for shadowy habitats. Travelers near troll domains often report an unnerving sense of being watched or hearing heavy footsteps, followed by guttural growls or deep vocalizations. Trolls are said to be fiercely territorial, attacking those who trespass in their areas. Some accounts suggest that trolls have rudimentary intelligence, using tools or constructing crude shelters near their lairs.

Additional Facts

- Some cryptozoologists theorize that trolls could explain mysterious disappearances in Norway's remote wilderness.
- Trolls are believed to be omnivorous, scavenging food or hunting small animals, and in rare cases, stealing livestock.
- The legends surrounding trolls may have developed as explanations for strange geological formations, unusual animal behavior, or eerie nighttime sounds.

Additional Resources

1. Erik Bjornsen, Cryptids of the North: Trolls and Other Mysteries.
2. Arvid Lund, Hominids in Hiding: A Study of Cryptid Species in Scandinavia.
3. Jorgen Fiske, Troll Territories: Investigating Cryptid Habitats in Norway.

93

WENDIGO

(Canada)

Description

The Wendigo is a terrifying cryptid from the folklore of Indigenous peoples in Canada, particularly the Algonquin, Cree, and Ojibwe nations. It is often described as a gaunt, humanoid creature with grotesque features: grayish, decaying skin stretched tightly over its bones, glowing sunken eyes, and sharp, jagged teeth. In some accounts, the Wendigo is said to have deer-like antlers or a skull-like head, adding to its monstrous appearance. Towering between 7 and 15 feet tall, it exudes an aura of insatiable hunger and malevolence. The Wendigo is believed to stalk remote forests and snowy wilderness, luring victims with eerie calls or mimicry before devouring them.

Origin

The Wendigo legend originates from the oral traditions of Indigenous cultures in Canada and the northern United States. It is traditionally seen as a symbol of greed, gluttony, and the horrifying consequences of breaking taboos, particularly those related to cannibalism. During harsh winters, when food was scarce, tales of the Wendigo served as cautionary warnings against resorting to inhuman acts for survival. Over time, the Wendigo became associated with mysterious disappearances and malevolent forces in remote, snowy regions.

Lore

The Wendigo is described as a creature cursed with eternal hunger, no matter how much it consumes. Legends suggest it was once human, transformed by acts of greed or cannibalism into a monstrous being. Encounters often include eerie howls, strange footprints in the snow, or the unsettling mimicry of human voices calling for help. The Wendigo is said to bring a sudden drop in temperature or an unnatural chill in the air, signaling its approach. Some stories claim the creature can possess individuals, transforming them into Wendigos through its corrupting influence.

Additional Facts
- The Wendigo is linked to Wendigo psychosis, a cultural phenomenon in which individuals develop a fear of turning into the creature, often accompanied by paranoia and cravings for human flesh.
- It is believed that a Wendigo can be destroyed by burning its icy heart or through sacred rituals performed by Indigenous healers.
- The Wendigo has inspired numerous adaptations in literature, film, and television, symbolizing unchecked greed and the primal fear of isolation.

Additional Resources

1. Basil H. Johnston, The Manitous: The Spiritual World of the Ojibway.
2. Howard Norman, Northern Tales: Stories from the Native Peoples of the Arctic and Subarctic.

94

YETI

(Himalayas)

Description

The Yeti, often called the "Abominable Snowman," is a cryptid believed to roam the icy and remote regions of the Himalayas. Typically described as a large, ape-like creature, the Yeti is covered in thick, dark brown or white fur, which allows it to blend into its snowy surroundings. Standing between 6 and 10 feet tall, it is said to have broad shoulders, long arms, and large, human-like feet that leave distinct tracks in the snow. Its face is often described as a mix of human and primate features, with deep-set eyes and a flat nose. The Yeti is known for its elusive nature, avoiding human contact while leaving behind evidence such as footprints, strange sounds, and fleeting sightings.

Origin

Himalayan folklore among the Sherpa and Tibetan peoples describes the Yeti as a spiritual being, revered as a guardian of the mountains and a symbol of the natural world's power. The creature is often linked to sacred sites and serves as a warning of the dangers in the harsh mountain environment. Western interest in the Yeti began in the 19th century, fueled by explorers and mountaineers reporting unusual footprints and other strange phenomena in the snow. Over time, the Yeti evolved from a spiritual figure in local traditions to a cryptid that captivates adventurers and researchers worldwide.

Lore

According to legend, the Yeti inhabits high-altitude forests and alpine tundra, moving stealthily through the treacherous terrain. It is said to possess immense strength, capable of moving large rocks and trees with ease. Some stories describe its eerie howls or growls, which echo across the mountains on cold, quiet nights. While generally shy and reclusive, the Yeti is feared as a territorial creature that may attack those who disrespect its domain. Tales of the Yeti often highlight the sacred and untamed nature of the Himalayan wilderness, serving as a reminder to respect its mysteries.

Additional Facts

- In 1951, British mountaineer Eric Shipton captured photographs of large footprints in the snow near Mount Everest, sparking global fascination.

- Despite numerous expeditions and findings like footprints and hair samples, no definitive evidence of the Yeti's existence has been discovered.

Additional Resources

1. Reinhold Messner, My Quest for the Yeti: Confronting the Himalayas' Deepest Mystery.

2. Daniel Taylor, Yeti: The Ecology of a Mystery.

About the Author

Ethan Howard is the founder of *A Blind Guy's View*, a YouTube channel where he reviews movies, with a particular focus on classic films. In addition to his work as a content creator, Ethan is also the proud father of two beautiful children, Hadley and Holden. He has written several books in his "Epic" series and is set to release his upcoming novel, *Eppica's Lily*, in late 2024.

Growing up in southern Oregon, Ethan has spent over 15 years working for the U.S. Forest Service and has taught Geospatial Science at Oregon Tech for more than a decade. Diagnosed with Stargardt's Dystrophy at the age of 14, Ethan has never let his visual impairment hold him back, demonstrating a deep passion for storytelling, the outdoors, and education. He hopes his resilience and creativity continue to inspire those around him.

Contact Ethan directly at ABlindGuysView@gmail.com

Soli Deo Gloria

Made in the USA
Columbia, SC
02 April 2025

56039073R00115